ISBN 978-0-267-84422-7
PIBN 10262630

THE ILLUSTRATED
OPTICAL MANUAL

OR

HANDBOOK OF INSTRUCTIONS
FOR THE GUIDANCE OF SURGEONS IN
TESTING QUALITY AND RANGE OF VISION, AND
IN DISTINGUISHING AND DEALING WITH
OPTICAL DEFECTS IN GENERAL.

BY

SURGEON-GENERAL SIR T. LONGMORE, C.B., F.R.C.S.

HONORARY SURGEON TO THE QUEEN;
PROFESSOR OF MILITARY SURGERY AT THE ARMY MEDICAL SCHOOL;
OFFICER OF THE LEGION OF HONOUR;
ASSOCIATE OF THE SOCIETY OF SURGERY OF PARIS;
CORRESPONDING MEMBER OF THE ACADEMY OF MEDICINE OF FRANCE;
ETC.

FOURTH EDITION,

ENLARGED, AND ILLUSTRATED BY 74 FIGURES FROM DRAWINGS AND DIAGRAMS

BY INSPECTOR-GENERAL DR. MACDONALD, R.N., F.R.S., &c.

LONDON :

LONGMANS, GREEN, AND CO.

AND NEW YORK : 15 EAST 16th STREET.

1888.

PREFACE

TO

THE FOURTH EDITION.

I HAVE endeavoured to make the present edition of the Optical Manual as complete a guide as possible to the diagnosis and management of optical defects, in the hope that it may prove useful as a text-book on the subject for civil as well as for military surgeons. In the former editions the necessities of military practice were almost exclusively kept in view, but in the present revised and illustrated edition, while bringing the text up to date in respect to the visual requirements and regulations of the military, naval, and other public services of Great Britain and India, I have added very materially to the work in order to try, in addition, to meet the wants of general practice. The explanations given in the several chapters into which the book is divided are expressed in as concise and simple terms as appeared to me compatible with a sufficient elucidation of the matters treated on in them, having regard to the needs of those who have not previously given particular attention to the branch of ophthalmic practice with which the work deals, and not forgetting the little time which, as a rule, is at the disposal of practitioners for the study of any special subjects beyond those which the demands of ordinary practice render essential. I have found it convenient to reprint a portion of the Preface to the third edition of this Manual, in order to show the aim and scope of those parts of the work which especially bear on optical practice in the public services. The advantages that are afforded by the illustrations, which appear for the first time in this edition, and, with the single exception of fig. 65, are taken from original drawings furnished by my late colleague, Inspector-General Dr. Macdonald, R.N., F.R.S., will be sufficiently apparent on observation of the illustrations themselves.

EXTRACT

THE PREFACE TO THE THIRD EDITION.

THE circumstances under which I was originally led to put forth this Manual of Instructions for testing and dealing with the various conditions of vision liable to be met with in persons seeking employment, or already engaged, in military service, have been explained in the two previous editions of this work, which were published in the years 1862 and 1874, and do not require repetition. Since the second edition of the Manual was published, so great have been the changes in some of the practical parts of optical manipulation, and such advances have been made in respect to length of range, and capacity for accurate fire, in the weapons with which soldiers now have to deal, that many passages of the Manual of that date have become obsolete, and the present edition has had to be less revised than re-written.

When the increased and increasing importance attached to the expert use of firearms of all descriptions, both rifles and guns, at very long ranges is remembered, it may reasonably be expected that before long even greater attention will be given by all persons in this country who are engaged in military pursuits to questions of quality of eyesight, and that more information will be demanded on the subject from medical officers than has hitherto been required from them. In the army, the firearm with which the infantry soldier has to become familiar is his rifle, and this he can only use with thorough efficiency when he has visual power enough to enable him to see clearly the objects at which he is required to aim, and to judge accurately their distances, whatever the range over which a projectile fired from the rifle may have to travel in order to reach them. A great deal of attention has been given of late years to improving the modes of instruction in musketry practice, and very recently important changes have been made in it, with a view to insure perfection, not so much in hitting a definite mark on a target, as to ensure accuracy of aim under conditions similar to those which are likely to occur when the soldier is engaged actively in the field; but, whatever may be the mode of instruction, so long as the rifle is such as it is, and objects of limited sizes, such as men, are to be fired at from distances of eight or nine hundred yards and upwards, an adequate power of eyesight must evidently be the prime ingredient necessary to insure the success of the marksman. The more this fact is appreciated, the more the importance will be felt of giving attention to the subject of the quality of eyesight of every one who aspires to effective employment of a rifle. It hardly seems too much to anticipate that, in respect to army service, at some future time the visual quality of every recruit will be as much recorded on his entry into the army as his height, chest measurement, weight, or any other of his physical conditions which are now registered; for some of them, considering all the sanitary precautions and personal care that are now taken to preserve the physical efficiency and good health of the soldier, have lost much of the importance that belonged to them in former days, and can hardly be regarded as equal in value to the amount of visual power which the man possesses, so far as his usefulness during the period he is engaged on active service is concerned. Military efficiency, the personal safety of troops, and economy of expenditure of ammunition are all involved in the capacity of soldiers for making an accurate use of the firearms placed in their hands. It is certain that the capabilities of the rifle can only be completely turned to account by persons who possess normal acuteness of vision, at least as regards the right eye; and it seems to be manifestly important, therefore, that the qualifications of each man who is destined to be a rifleman should be thoroughly known on his starting in the service, so that, on the one hand, the time and efforts of instructors may not be wasted in trying to teach men matters which from natural causes they may be totally incompetent to acquire, whatever labour may be devoted by themselves or others to the attempt, and also, on the other hand, that the men may be distinguished, and made known to commanding officers, who possess the necessary optical qualities for becoming sure and reliable marksmen.

Not improbably, as further advances are made in musketry instruction, a greater influence will be exerted by physiological optics on certain parts of the teaching. The objects painted on targets, and employed in 'educating' men in the use of the rifle at the various range practices, have not been designed, as shown in the text, on a uniform visual standard, but seem rather to have been settled, especially as regards their dimensions at the various distances at which they are usually placed for firing practice, according to the proportions which have been considered, from personal observation, to be the most suitable for marks to be aimed at. There is apparently no sufficient reason why all such objects should not be fashioned on a regularly graduated scale of dimensions and configuration in relation to distance, and be in exact accordance with the optical conditions, so far as the objects themselves are concerned, under which they would present themselves as marks to be fired at in actual warfare. If a series of objects on such principles should be brought into use, the quality of sight necessary for a satisfactory execution of the contemplated task at any particular range of practice could be defined with almost mathematical precision.

Medical officers at present only have to determine the question whether a man is optically fit for military service so far as the possession of a set minimum standard of vision is concerned; but in performing the duties of recruiting, they may have in the future to answer several questions of a more complex kind. They may be required to furnish information on such questions as the following: Is the man visually qualified to become a marksman up to the longest range for which the rifle is capable of adjustment? If not fit for a complete marksman, up to which class of practice does his visual power admit of his being advantageously trained? If not fit for the use of an arm of precision in the first line of the army, is he fit for duty in the ranks of the Militia or Volunteer forces? If not fit for the duties of a rifleman, is he visually qualified for service in the Commissariat and Transport, or for any other corps or department of the army? After a sufficient number of records on these subjects has been accumulated, a conclusion may be arrived at on certain questions, which are regarded under different aspects in different armies, and which may well admit of different solutions in different countries; as, for example, whether the proportion of men in this country, whose sharpness of sight is inferior to the normal standard owing to refractive defects, is so great as to render it advisable, from a military and financial point of view, to allow correcting spectacles to be used in the ranks of the army? and, in case of a decision being come to that it is advisable, whether the permission to wear them should be restricted to spectacles of certain descriptions, and, if so, of what descriptions?

I do not think it too much to assert that an acquaintance with the subjects described in this Manual, combined with a moderate amount of practice, will enable medical officers to furnish satisfactory information, when required, on the various points to which I have alluded, as well as on any others of a similar nature that may arise, and to carry out any orders that may be issued in respect to visual examination, or to correction of ocular defects among the officers and men of the army, in all ordinary cases which depend on faults of refraction or accommodation. At the same time it should not be forgotten that under the usual circumstances of service, in consequence of the multitudinous duties which devolve on medical officers, it would be too much to expect that more than a limited number among them will find the time or opportunities for becoming experts in ocular investigations. Arrangements will probably still have to be made, as hitherto, for complicated and doubtful cases of defective vision to be sent to general hospitals, and referred to medical officers who have acquired a particular acquaintance with the visual conditions which are liable to be encountered, by having had the means of carrying out extended observations of them at such institutions, and where also there will generally be the opportunity of using special optical appliances which cannot be expected to be found at more limited establishments.

Although the metrical system of numeration of lenses and of measurement in general is now ordinarily employed by ophthalmic surgeons, and is no doubt destined to supersede the duodecimal system everywhere, there are still many practical difficulties in the way of its adoption by British military medical officers in the different parts of the world in which they have to perform their duties. They have not, as a rule, cases of lenses numbered in dioptrics available for their use; the ordinary appliances for measurement at their command are divided into inches; and these will probably remain the conditions in respect to such matters until the metre becomes the standard of measurement for the ordinary purposes of society and commerce. It thus becomes necessary for British medical officers to be acquainted with both systems of measurement, and to be able to convert any expressions according to the metrical system which they may happen to meet with, into their relative values on the duodecimal system. The means of doing this are fully explained in this Manual, and where references are made to optical measurements in the body of the work, they are usually stated in figures belonging to both the metrical and duodecimal systems of measurement.

The present work is limited to a study, theoretical and practical, of those varieties of the visual function, which for the most part are independent of morbid processes, and in considering these conditions of sight, their bearing on military service is always kept in view. I have at the same time added short explanations on a variety of optical matters more or less directly connected with visual examination and the correction of visual defects, a knowledge of which is essential to a right understanding of the principles on which the practical part of the work is conducted. Experience in teaching the modes of conducting the visual examination of recruits and soldiers, and the practical correction of visual defects, has proved to me the need of such information being given, and I hope that some of the explanations and matters of fact, which it has been found necessary to impart in the course of instruction at the Army Medical School, may prove to be serviceable as memoranda to medical officers in the larger sphere of the Army Medical Department itself.

CONTENTS.

CHAPTER I.

OPTICAL MEMORANDA : SIGHT EXERCISES AT MUSKETRY INSTRUCTION :
DESCRIPTION OF LENSES, &c.

CHAPTER II.

ON THE VARIOUS CONDITIONS OF SIGHT WHICH RESULT FROM DIFFERENCES IN REFRACTIVE POWER AND FOCAL ADJUSTMENT OF THE EYE.

CHAPTER III.

OBJECTIVE ASSESSMENT OF OCULAR REFRACTION.

CHAPTER IV.

ACCOMMODATORY FUNCTION OF THE EYE.

CHAPTER V.

ON IMPAIRED VISION CONNECTED WITH STRABISMUS.

CHAPTER VI.

ON DEFECTS OF COLOUR SENSE.

CHAPTER VII.

ON VISUAL ACUTENESS AND PARTICULAR VARIETIES OF WEAK AND IMPAIRED VISION.

CHAPTER VIII.

ON THE POWER OF SIGHT REQUIRED FOR RECRUITS, WITH A DESCRIPTION OF THE TEST-DOTS EMPLOYED FOR TESTING IT.

CHAPTER IX.

ON THE MANNER OF CONDUCTING THE VISUAL EXAMINATION OF RECRUITS
AND SOLDIERS, TOGETHER WITH AN ACCOUNT OF THE QUALITIES OF
SIGHT WHICH DETERMINE THE SELECTION OF OFFICERS AND MEN FOR
ARMY, NAVY, AND INDIAN GOVERNMENT SERVICES.

CHAPTER X.

APPENDIX OF EXTRACTS FROM MUSKETRY AND OTHER REGULATIONS CONCERNING EYESIGHT WHICH AFFECT MEDICAL OFFICERS, TOGETHER WITH NOTES ON CERTAIN OTHER MATTERS REFERRED TO IN THE BODY OF THE WORK.

ILLUSTRATIONS.

OPTICAL MANUAL.

CHAPTER I.

Preliminary Remarks.—The chief purpose of this manual is to
furnish surgeons with such information as will assist them in
ascertaining the quality and power of vision of any persons who
may be submitted to them for visual examination, and also, that
will enable them to pronounce an opinion on the fitness, so far
as sight is concerned, of recruits for the army and candidates for
appointments in the various public services of the country. It is
necessary, therefore, to describe the various qualities of vision which
are liable to be met with. The chief characteristics of the different
conditions of sight depending upon differences in the refractive
power and conformation of the eye, and the modes of determining
the degrees in which these differences exist, will be treated on in the
second chapter; while other states of vision, either associated with
them, or occasionally consequent upon them, will be remarked
upon in succeeding chapters. The visual needs of men employed
in the different branches of the military services, army, navy,

B

militia, and volunteers, and also of the Indian service, the regulations concerning the degrees of defective vision which disqualify men for occupation in them, and the means to be employed for estimating whether they possess the requisite visual qualifications, will be fully described subsequently.

In carrying out the practical instruction of young surgeons on the optical examination of the eye at the Army Medical School as applied to the circumstances of military service, it has been constantly found necessary to explain various elementary matters on optics which must be known before the phenomena of vision, whether normal or abnormal, or the principles on which refractive and accommodatory defects of vision are corrected, or the means by which the correction is effected, can be properly understood. Several points connected with the special training which all men undergo who have to acquire a technical acquaintance with the use of arms of precision, such as the practice of judging distances of objects, of aiming at long ranges, and other such exercises in which quality of vision exerts a material influence, also require some preliminary explanations to be given to those who have not previously had to consider such subjects. The chief of these optical memoranda and allied topics are given in this introductory chapter in as plain and concise a form as possible for convenience of reference.

Vision.—Normal vision exists when, firstly, each eye is so constructed that a sharp and exact image of the object toward which the eye is directed is impressed on its retina ; and when, secondly, a correct appreciation of the form, distance, colour, and most of the physical qualities of the object so depicted is produced in the mind of the observer. To effect the first of these conditions, the curvatures, refractive qualities, and mutual relations of the transparent media of each eye must be such that the rays of light proceeding from the external object are all brought by them to suitable foci upon the proper nerve elements of the retinæ. To effect the second of these conditions the retinæ, the optic nerves and their cerebral connexions must be anatomically and physiologically healthy and fully developed ; the images of objects must be of sufficient size, sufficiently but not excessively illuminated, and must remain sufficiently long on the retinæ; the mutual relations of accommodation to distance and of direction of the visual lines must be in normal accord ; and the perceptive faculties must have been duly educated.

Monocular Vision.—Vision by one eye. An object, though solid, when looked at by one eye, the other being closed, appears as a plane figure, having two dimensions, viz., length and breadth (see fig. 1). When a group of objects is regarded monocularly, although the lights and shadows among them are visible, no means are afforded of determining their relative distances or depths. They are presented to the view like a picture on a flat surface.

These defects of monocular vision may to a certain extent be counteracted by long practice, familiarity with objects and their physical qualities, and acquaintance with the rules of perspective, but the effects of binocular vision can never be fully realised by the use of one eye alone.

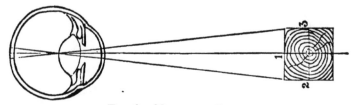

FIG. 1.—MONOCULAR VISION.

The superficies, 1, alone visible ; 2 and 3 out of range.

Binocular Vision.—Single vision by two eyes. The visual lines, or lines prolonged from the fovea centralis and passing through the nodal point of each eye, meet in the same point of an object. If the object be solid, the two images of it, or rather such portions of them as are common to the two eyes, are impressed upon corresponding parts of the two retinæ. Images of certain parts of a solid object are only formed in each eye singly—in the eye on the same side as the parts concerned. Under these conditions the images formed in the two eyes respectively are mentally combined,

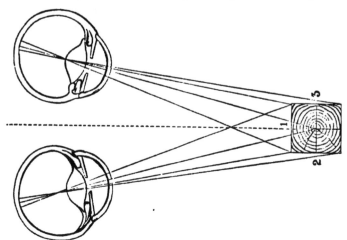

FIG. 2.—BINOCULAR VISION.

The surfaces, 1, 2, and 3, being visible, determine the perception of relief with three diameters.

and produce sensorially the effect of having proceeded from a single retinal impression. So far, then, as concerns visual perception, notwithstanding the dual sensation, the object appears, as it really is, single ; having three dimensions, viz. length, breadth, and thickness ; and consciousness of *relief* is obtained (see fig. 2).

A wider field and range of vision are obtained with two eyes than with a single eye. The double impression on the two retinæ in binocular vision also causes the perception of objects to be more vivid than it is in monocular vision. When a coloured object is

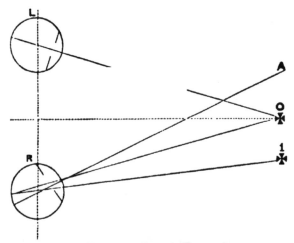

Fig. 3.—Diplopia. Images Homonymous.

R, right eye; L, left eye; A, optic axis; O, object; I, referred image.

seen by two eyes its brightness appears to be greater and its tint more saturated than when it is looked at by one eye singly. Owing to the visual perception being thus intensified, the rapidity

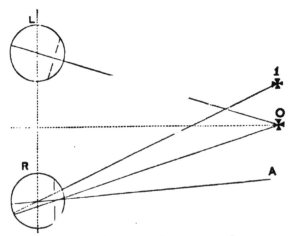

Fig. 4.—Diplopia. Images crossed.

R, right eye; L, left eye; A, optic axis; O, object; I, referred image.

with which objects are recognised, or visual alertness, is increased under binocular vision. The greater or less degree of convergence of the visual lines which accompanies binocular vision also helps the observer to form ideas of the distances at which the different objects looked at are situated. It is obvious that, as regards near objects which are situated in the median plane between the two eyes of an observer, the convergence of the two eyes must be in exact accord in order to ensure perfect singleness of vision of these objects.

Diplopia.—Double vision. This results when an object is perceived by both eyes, but its two images are not formed on corresponding parts of the two retinæ, so that the images cannot blend for the production of a single sensory impression. The visual lines of the two eyes do not meet and join each other in the object. When the separate images correspond in their relative positions with the relative positions of the two eyes, they are described as *homonymous*; when, on the contrary, the image to the left in position belongs to the right eye, and that to the right is the image perceived by the left eye, they are spoken of as *crossed images* (see figs. 3 and 4).

Optic Axis.—A line prolonged from the centre of the cornea,

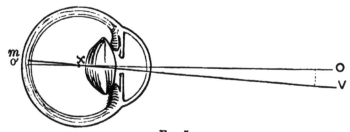

Fig. 5.

oo′, optic axis ; x, nodal point ; vm, visual line ; m. macu'a lutea ; oxv, angle formed by line of optic axis and visual line.

through the nodal point of the eye, to the retina. The posterior pole of the optic axis impinges on the retina a little to the inner side of the fovea centralis (see fig. 5).

Visual Axis, or visual line.—The line along which the axial ray proceeding from the point of the object looked at passes to terminate in the fovea centralis. In the emmetropic eye it enters a little to the inner side of the centre of the cornea, crosses the optic axis at the nodal point, and falls on the fovea centralis ; or *vice versâ*, starts from the fovea centralis, and, passing through the nodal point, terminates in the object looked at. The visual line thus forms an angle with the optic axis, the apex of which is at the nodal point of the eye ; the size of this angle varies in different eyes according to their refractive qualities, whether emmetropic, myopic, or hypermetropic.

In binocular vision, when the object which is viewed is remote, the visual lines of the two eyes are parallel, or nearly parallel, with each other ; in proportion as the object viewed approaches nearer to the observer, the visual lines from the two eyes necessarily become more and more convergent in direction.

Radiation of Light.—The expression *ray of light* indicates the straight line along which light progresses, and the term *radiation* signifies the emission of rays from a source of light. When the luminous body is central, it emits rays of light in straight lines from all its points and in all directions (see fig. 6). A collection of rays emitted from a luminous point, and circumferentially so limited in their passage as to assume a conical outline, is called a *diverging pencil of rays*; and the apex of the cone from which they proceed is called the *focus* of the pencil. When the divergency of rays proceeding from a remote point of light is inappreciable, the rays are described as *parallel rays*. When rays are artificially

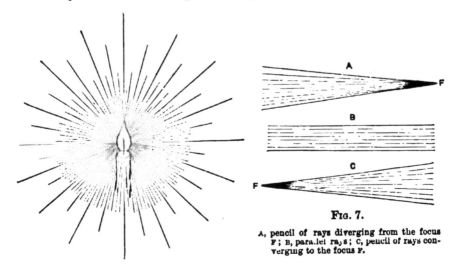

FIG. 7.

A, pencil of rays diverging from the focus
F; B, parallel rays; C, pencil of rays converging to the focus F.

FIG. 6.—RADIATION FROM A CENTRAL
SOURCE OF LIGHT.

caused to converge to a common point, they are together spoken of as a *converging pencil of rays*; and the point at which they all meet is called the focus (see A, B, C, fig. 7).

The illuminated surface of any object, in respect to the light it is receiving, is regarded as lighted by the bases of a number of divergent pencils of rays, whose foci are at the source of illumination; and, in respect to the light it is imparting by reflection from its surface, by means of which it is rendered visible, it is regarded as composed of an infinite number of luminous points, which points are the foci of a corresponding number of diverging pencils of rays (see fig. 8). It is important, when speaking of the rays proceeding from an illuminated object, not to confound the rays emanating from two points remote from each other (as from the extreme points between which the visual angle is included, for example) with the

rays proceeding from each luminous point of the object independently.

Permeability of Bodies to Light.—All substances are probably
pervious to light in different degrees and to different depths. If
the substance be one which allows no rays of light to pass through
it, it is said to be *opaque*; if it be one which admits of light
passing through its substance, or transmits rays sufficient to allow
objects to be seen through it, it is termed *transparent*; if the substance be one like thin porcelain, which permits the passage of
light through it, but does not allow objects to be seen through it, it
is described as being *translucent. Shadows* result from opaque
bodies intercepting the passage of light. All bodies, whether opaque
or transparent, vary in aspect and colour according as they vary
in reflecting, absorbing, or decomposing the light which falls on them.

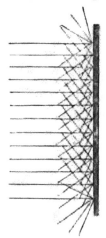
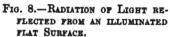

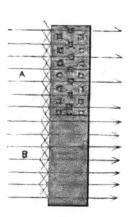

FIG. 8.—RADIATION OF LIGHT RE-
FLECTED FROM AN ILLUMINATED
FLAT SURFACE.

FIG. 9.—THEORETICAL ILLUSTRATION OF
VARIOUS DEGREES OF PERMEABILITY
TO LIGHT.

Reflection of Light.—All bodies, whether opaque or transparent,
with smoothly-polished surfaces, under certain conditions turn
away the rays of light which fall on them from their original
direction. This is termed *reflection of light.* There are two
general laws of reflection of light. The first is that whatever may
be the angle which a ray impinging on a polished surface forms
with the normal, or perpendicular, to that surface, the angle at
which it is turned away from that normal will be the same; in
other words the *angle of reflection* is equal to the *angle of incidence*,
and on the opposite side of the normal (see fig. 10). The second
law is that the plane in which the incident ray is found will be the
same as the plane in which the reflected ray is found, or in other
words, the *plane of incidence* coincides with the *plane of reflection.*
All the phenomena of reflection of rays of light from polished sur-

·faces, whether plane or curved, take place in accordance with these laws. If the polished surface be either level or have a regular curvature, the reflected rays of light produce images of the objects from which the rays have proceeded. If the surface be roughened,

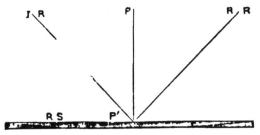

FIG. 10.—REFLECTION OF LIGHT.

PP′, perpendicular, or normal, to RS, the reflecting surface; IR, incident ray; RR. reflected ray; PP′R, the angle of reflection, is equal to PP′I, the angle of incidence, and is on the opposite side of the normal, PP′.

there will still be reflection of light, but the reflected rays are irregularly dispersed, and no images are produced.

Refraction of Light.—Rays of light proceed in straight lines so long as the medium through which they are travelling is of uniform density. When a ray passes obliquely from a rarer into a

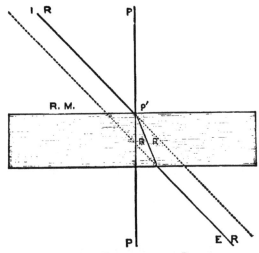

FIG. 11.—REFRACTION OF LIGHT.

PP′P, perpendicular, or normal, to RM, one surface of a refracting medium, such as a piece of glass with opposite parallel surfaces; IR, incident ray in air; RR, incident ray refracted towards the perpendicular in passing through RM; RR, emergent ray refracted from the perpendicular in passing out of RM into air.

denser medium, it is bent or *refracted towards* a line drawn perpendicularly to the surface of this medium at the point of incidence; conversely, on passing obliquely from a denser into a rarer medium, it is *refracted from* a line drawn perpendicularly to its surface (see fig. 11). This change of direction commences at the surface of separation of the two media. When a ray of light passes through media of different densities *perpendicularly* to the surfaces where these media are in contact with one another, the ray travels onwards in one and the same straight line. In accordance with this rule, a ray of light passing through air and impinging on the surface of a piece of glass perpendicularly to the point of incidence, passes on unchanged in direction ; but if it fall on the surface with a slanting direction, it is refracted in passing through the glass towards the perpendicular to the surface at its point of incidence. Any substance, whether liquid or solid, through which light can pass, will produce a similar effect to that produced by the glass, but the refraction will vary in degree in substances of different kinds. The deviation of rays of light from their original direction on passing obliquely from one into another medium of different density takes place according to fixed laws, and the investigation of these laws, and of the phenomena which result from them, constitute the

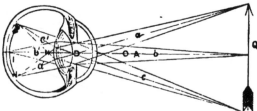

FIG. 12.—INVERSION OF RETINAL IMAGES.

Q, object ; a, b, and c, upper, middle, and lower diverging pencils of rays emanating from Q ; OA, optic axis ; D, centre of pupil, or decussating point of the central axes of all the cones, from whence they continue to diverge, while the marginal rays converge towards them (as seen at a', b', and c'), and finally depict the inverted image (i) on the retina.

branch of optics generally termed *dioptrics*. The three following laws are constant in all cases of refraction :—(1.) The angle formed by an incident ray of light with the perpendicular to the surface, *or the angle of incidence*, and the angle formed by the refracted ray with the perpendicular, *or the angle of refraction*, are in the same plane. (2.) The incident ray and the refracted ray are always on opposite sides of the perpendicular. (3.) Whatever the inclination of the incident ray to the surface, the sine of the angle of incidence has a constant ratio to the sine of the angle of refraction. These laws apply to curved surfaces equally with plane surfaces, and hence, when the form of surface and nature of a refracting medium are known, the path of any refracted ray can always be determined.

Law of Visible Direction.—Each point of an object is seen in a line perpendicular, or nearly so, to the point of the retina which its image impinges.

Inversion of Images on the Retina.—The pictures formed on the retina of external objects are *inverted* and *curved*, owing to the action of the optical apparatus of the eye, together with the concave form of the retinal receiving surface (see fig. 12). The mind, however, does not judge of the positions of objects, whether primary or reflected, according to the part of the retina on which their images happen to fall ; if it did, the positions of the things would appear to change with changes in the position of the eyes looking at them. But the mind judges of the positions of objects by following, as it were, the directions of the axial rays proceeding to all the points of these objects from the parts of the retina on which the corresponding images of such points are pictured. Hence, though the images of objects looked at directly are inverted on the retina by the action of the refracting media of the eye, the mind, following the lines of light to their sources in accordance with the law of visible direction, sees them in their true positions. The images of objects below the level of the visual diameter are pictured in the upper retinal hemisphere ; the images of objects above this

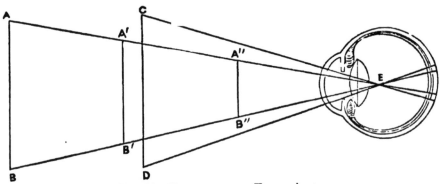

FIG. 13.—EXPLANATION OF VISUAL ANGLE.

Visual angles ABE, CDE, &c. CD, nearer to the eye than AB, though alike in size, has a larger visual angle and forms a larger retinal image. A″B″, A′B′, AB, though smaller than each other, have the same sized visual angle, and form images of the same size upon the retina. Distance and size together determine the magnitude of the visual angle, the size of the retinal image, and the apparent sizes of objects as seen by the observer.

line in the lower retinal hemisphere ; they are equally reversed in the lateral portions of the retinal picture ; but, nevertheless, all the objects are seen in their real positions and relations to each other. This equally applies to the reflected images of real objects : the reflected images are inverted on the retina, but they are seen truly in the forms in which they proceed from the reflecting surface.

Visual Angle.—The visual angle is the angle included between two rays proceeding from the opposite extreme limits of an object looked at by the eye and meeting at a point within the eye. These rays, having met, cross each other and pass onward to assist in forming the image on the retina. The point at which they meet is known as 'the point of intersection,' or 'nodal point,' of the eye. The size of the visual angle depends on the linear dimensions of an object, and on the distance of the object from the eye. If the object be of a fixed size, the size of the angle under which it is seen will vary inversely as its distance from the eye ; if the distance be fixed, the size of the visual angle will vary in a *direct* ratio with the size of the object. The angle is similar on each side of the point of intersection—towards the object, the 'visual angle,' and towards the image of it on the retina, the 'retinal angle.' The expression that an object occupies so many degrees in the circumference of a circle of which the eye is the centre, or that it subtends an angle of so many degrees, has the same significance as 'the size of the visual angle of the object.'

The size of the image of an object formed on the retina varies as the retinal angle, and therefore as the visual angle varies under which the object is seen. The larger the visual angle, the larger the retinal image. If the size of an object remain the same, the visual angle it subtends is increased in proportion as it is brought nearer to the eye ; and hence, the frequently observed approximation of printed letters by Hc. patients to their eyes, although the diffusion of rays about the retinal images and the strain on the accommodation are increased by the proceeding. A similar approximation of print generally takes place among amblyopic subjects. In both instances the retinal images are increased in size, and the area of sentient visual impression proportionally enlarged ; and the advantages attending these results preponderate over the disadvantage of the loss of distinctness of outline, due to the diffusion of the marginal rays.

The retinal image will also vary in size as the distance from the point of intersection of the rays forming it to the plane of the retina varies. In a short, or hypermetropic eye, the distance from the nodal point to the retina will be less than it is in an elongated or myopic eye, and the rays proceeding to form the retinal image of an object, being sectionally interrupted in their course earlier in the former than in the latter case, the areal size of the image will be necessarily less in the former than it is in the latter instance.

If the position of the point of intersection in an eye is made to alter, the size of the retinal image will be altered also, but this can hardly happen except artificially by placing a convex or concave lens before the eye. If a convex lens be placed before an eye, the point of intersection will be caused to advance, and the retinal image will become enlarged : if a concave lens be similarly placed, the point of intersection will be caused to recede, and the retinal image will be lessened in size.

Field of Vision.—The term 'field of vision' signifies the whole of the space, including the objects comprised in it, which is perceptible to sight in one fixed position of the eye, or, in binocular vision, of the two eyes.

Monocular Field of Vision.—When a single eye is directly fixed in a given direction its horizontal limits of visual perception are comprised within an angle of about 123°.

Binocular Field of Vision.—When both eyes are fixed on an object situated at such a distance in front of the observer that the visual axes of the two eyes are practically parallel with one another, the limiting lines of visual perception or luminous impression on each side of the face form an angle of about 90°, so that the general field of vision of the two eyes under these conditions has a horizontal limit of about 180°. The precise limits of the field of vision will vary with the amount of projection, and other peculiarities in form of the facial features. A certain central portion of the visual field is common to both eyes ; the right and left temporal portions of the field are only proper to the right and left eyes respectively. The absence of this monocular part of the binocular field of vision is a serious drawback from the usefulness of a soldier in military service.

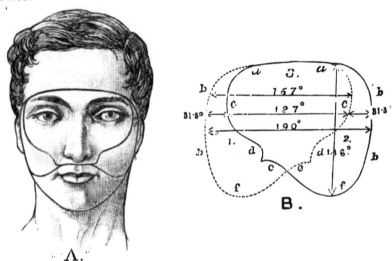

Fig. 14.—General, Common, and Proper Fields of Vision.

A. Sketch showing the visual limitations, in a fixed position of the face, due to the projection of the principal features.
B. Diagram showing the outlines of the common, proper, and general fields of vision, with measurements. a, b, f, b, outline of general field of vision ; 1 and 2, right and left proper fields of vision ; 3, common field of vision ; aa, upper limit due to frontal projection ; cc, limits due to malar projections ; ee, limit produced by labial projection ; dd, by nasal projection ; bb, limits due to projections of cheeks.

Measurement of the Monocular Field of Vision.—Instruments specially constructed for examining and providing the data for mapping out a field of vision are called *perimeters*. In the absence of these special instruments, the extent and shape of the field of vision of an eye may be obtained by causing the patient to look at a given point, and, while the eye is fixed upon it, drawing an outline of the boundary of distinct vision in all directions around it. The most convenient plan is to place the patient with one eye covered, and the other free, at a distance of about two feet in front of a black board placed perpendicularly to the line of sight. On this board, at a level with the eye under examination, a small cross in white chalk is drawn. The patient is desired to fix his eye on this mark. At the same time the surgeon, who must watch that the patient does not look away from the cross, holds a piece of chalk between his fingers of his right hand, and carries it from point to point over the board by slight quick movements of the hand, jotting down the points in various directions where it ceases to be seen. These points are now joined together by lines, and thus an outline of the shape of the field of vision is formed. If the map thus made be copied on paper, it can be retained for comparison with other diagrams to be made in a similar way on future occasions.

If it be important to examine the field of vision with greater precision, separate outlines can be obtained by a similar plan showing where vision sufficiently distinct to count fingers ceases, and more externally where imperfect vision or luminous impression ceases and complete absence of sight begins. Any loss of visual function in particular spots of the retina may also be traced and noted during the examination by moving the chalk slowly and carefully over the field, and directing the patient to mention whenever it disappears from view altogether or is only seen obscurely. Particular irregularities of form, and limitations in extent, of the field of vision will often be rendered manifest by this mode of examination in cases of weak and defective vision depending on disorders of the optic nerve and retina.

When it is not required to make a picture of the field of vision but only to ascertain quickly the extent of the field, or whether there is any break in it in any given direction, the following method by the finger will answer the purpose readily and quickly. The surgeon, standing about two feet off, and face to face with the patient, desires him to close one eye, the right, for example, and at the same time closes his own left eye opposite to it. He now desires the patient to look steadily into his right eye, and while the two eyes, the one of the surgeon and the other of the patient, are thus directed to each other, the surgeon moves his forefinger from a central point midway between his own eye and that of the eye under observation, in all directions towards the limits of the field of vision. He is thus able to note the extent of the patient's range of vision, by comparing it with his own range. This ready method has the advantage of being capable of being put into execution anywhere without need of any appliances or previous preparation.

Movements of the Eye.—The external muscles connected with the globe of the eye cause it to revolve round an ideal centre—its centre of rotation. Consequently if the anterior aspect of the eye is caused to move in one direction, the posterior aspect of the eye will be caused to move in the opposite direction. The movements of the eye are commonly described with reference to the movements of the anterior pole of the optic axis or central point of the cornea. When the central point of the cornea accords with the point of intersection of one line drawn transversely between the apices of the two angles of the palpebral aperture, and another line drawn vertically midway between its two extremities, or does so very closely, allowance being made for the slight angle formed by the intersection of the optic axis and visual axis, the eye is said to be directed straight forwards; when the centre of the cornea is above or below the horizontal line just named, the eye is said to be directed upwards or downwards respectively; and if to the inner or outer side of the vertical line just named, it is said to be turned inwards or outwards respectively. It is in accordance with these distinctions of position that when strabismus exists the dis-

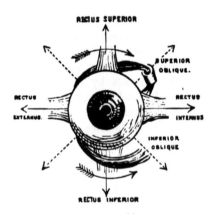

Fig. 15.—Ocular Movements.

The arrows show the direction of the movements. The four recti muscles singly effect purely vertical and transverse movements (plain arrows), while those in either diagonal line are produced by the joint action of two recti muscles (dotted arrows). Rotation to the right or left is effected by the oblique muscles (curved arrows).

placement of the deviated eye is said to be internal or convergent, external or divergent, and upwards or downwards. In normal binocular vision, when the two eyes are looking directly forwards at a distant object, the visual lines are practically parallel with one another, and perpendicular to an imaginary line joining the centres of rotation of the two eyes.

Field of View.—The term 'field of view' signifies the space over which objects can be seen clearly by an eye when it moves round its centre of rotation to the extreme limits of which it is capable, the head of the observer at the same time remaining stationary. In binocular vision, the 'field' of view signifies the visual range of the two eyes under like conditions.

Ocular Movements affecting the Field of View.—Fig. 15 shows the manner in which the visual range is obtained, and indicates the directions in which the movements affecting the field of view occur.

Normal Range of Motion of the Eye consistent with Vision.—I find that my own eye can turn through an angle of about 140° horizontally, and of 138° vertically, and perception of objects be retained. On my eye being directed to a point straight before it, it can turn from it 50° towards the nose and nearly 90° outwards in the horizontal plane, or through an angle of 48° from it upwards and 90° downwards. These measurements are not, however, universal; they are subject to variations according to individual circumstances, viz. to personal peculiarities in the shape and amount of projection of the parts near to the eye, as the nose, the margins of the orbit, the eyebrows, and other structures.

Objects on the Visual Field.—The place and space occupied by objects in the field of vision are measured by the visual angles under which they are seen. The *apparent size*, or lineal measure, of an object is estimated by the size of the visual angle alone. To estimate the *true size*, the distance of the object, as well as the size of visual angle, must be known. Conversely, the *true size* of an object being known, the visual angle enables us to form a judgment of the distance at which it is placed from us. The apparent size, or lineal measure, is to be distinguished from the apparent *superficial* size, or measure of surface of an object. The lineal measure, as before mentioned, varies inversely as the distance of the object; the measure of surface varies in proportion to the squares of the lineal measure at different distances.

Illumination of Objects.—The quantity of light received on a given object, or unit of surface, is reduced inversely as the square of the distance from the source of light to which the object is removed. Let a printed paragraph be placed at a certain distance from a lighted candle, and the paragraph be then removed to double the same distance from the light, the intensity of illumination of the paragraph in the more distant of the two positions will be only one-fourth of what it was in the nearer one; and the rule that applies to the whole paragraph equally applies to each letter of the print composing it.

Apparent Brightness of Objects at Different Distances in open Daylight.—The vividness of the light under which objects appear in open daylight at different distances does not, however, vary with the distances at which they are respectively placed, excepting so far as those which are more remote may be affected by the thicker stratum of air through which they are seen, provided that all the objects face in the same direction. Supposing the atmosphere to be perfectly clear and transparent, the apparent illumination of a series of targets, one more remote than the other but facing in the same direction, or of a line of men one in front of the other, will be exactly similar. A transverse section of the cone of rays entering the pupil from each illumined point of the farther object, and therefore the total quantity of light emitted by the whole object, will vary inversely as the square of the distance to which the object is removed, but so also will the apparent area of the object. If one object be placed at double the distance of the other, the area of its image on the retina will be reduced to one-fourth of that of the nearer one, and so also will be the amount of light entering the eye from the more remote object, so that the effect as regards apparent brightness will necessarily remain the same. The area of the image on the retina is reduced in the same proportion as the quantity of light which forms it.

In fig. 16 the targets are represented as being at equal distances apart, and, as may be seen by the nearer targets which admit of observation in the drawing, each is one-fourth smaller than the target in front of it, and one-fourth larger than the next target beyond. As the amount of light received by the eye from each target varies in the same proportion as the visual angle under

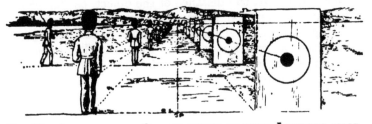

FIG. 16.—APPEARANCES OF OBJECTS PLACED AT EQUAL INTERVALS APART.

GL ground-line, on which the observer is supposed to stand ; SP, standing-point ; HL horizontal line on a level with the observer's eye, all round : PS, point of sight or position of observer's eye on the HL ; VL vanishing lines drawn from the salient or special points of all objects into the point of sight, determining their diminution in size as the distance increases. The observer is represented by the perpendicular.

which it is seen the apparent brightness of the object is unchanged. The same observation applies to the human figure under like circumstances.

Motion of Objects in the Field of Vision.—The movement of any visible object *across* the visual field, or the change of its position relatively to the positions of other objects, is, of course, accompanied by a similar movement of its image across the retinal picture. The movement which is thus rendered visible is called its *apparent motion*. If the object move in a direct line towards the centre of the eye, no change occurs in the position of the image of the object, and its movement is not apparent. The extent of apparent motion of an object is measured by its *angular motion* ; that is, by the angle formed by two lines drawn from the point of visible departure, and point of visible arrest of motion, of the object, to the point of intersection within the eye. The *real movement* of an object, in respect to direction and extent, is estimated by other means—by its change of apparent size as its distance varies, by its relations to other objects in the visual field as regards position, distance, distinctness, and other indications.

It is not to be forgotten that if the object moving across the field subtend too small a visual angle, or pass with too great rapidity, or is insufficiently illuminated, its movement may not be apparent; either because its image impinges on too minute a portion of the retinal surface to render it perceptible, or because there is not light enough to make an adequate visual impression, or because time enough is not given for the impression to be made. The movement of a gunshot travelling *across* the field may not be apparent, for the reasons just named; while, on the other hand, a similar shot or shell moving in a nearly direct line towards the eye can be seen because its image is sufficiently persistent, while, although it has no angular motion, its movement of approach may be inferred from increase of size or the louder sound proceeding from it in proportion to its increased nearness to the observer.

Infinite Rays.—This term, in reference to vision, is employed to express rays of light proceeding from an object and entering the eye in parallel lines, or rather in lines which are so nearly parallel with each other that their divergency is almost inappreciable; and the expression 'infinite distance,' or *infinity,* is used to signify in effect the distance from which incident rays possessing collectively such a parallel direction might have originally started. The angular measurement of that portion of the cone of rays proceeding from a luminous point which falls on the eye, so far as vision is concerned, is determined by the distance of the point from which the light radiates, and by the diameter of the transparent cornea, or rather by the distance apart of the opposite borders of that area of the cornea in connection with the pupil by which the passage of the rays into the interior of the eye is permitted. Practically, for optical purposes, the incident rays proceeding from every luminous point of an object about fifteen feet distant from the spectator are regarded as parallel rays, and, from this distance up to that of a fixed star, the rays reaching the eye are ordinarily spoken of as *infinite rays.* The angular divergency of a pencil of rays incident on the cornea from a luminous point at a distance of 15 feet from the eye, supposing the diameter of the cornea to be two-thirds of an inch, would be only about twelve minutes of a degree (12' 40"). But a considerable portion of the peripheral rays even of this slightly divergent pencil would be intercepted by the iris, so that the rays entering the pupil of the eye would be still more approximately parallel with the line of the visual axis. If the area of that portion of the cornea which receives the precise amount of rays that enter the pupil be taken to be one-eighth of an inch in diameter, the divergency of the rays impinging upon it from a luminous point at a distance of 15 feet would only amount to 2' 24".

Rays from objects nearer to the eye than fifteen feet are some-

C

times spoken of as *finite rays*, and the divergency of such rays increases in proportion to the proximity of objects to the eye.

On referring to fig. 17, it will be seen that the whole of the divergent pencil of rays proceeding from the luminous point F is

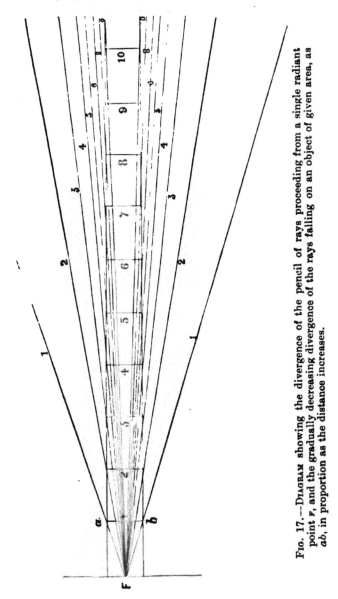

FIG. 17.—DIAGRAM showing the divergence of the pencil of rays proceeding from a single radiant point F, and the gradually decreasing divergence of the rays falling on an object of given area, as *ab*, in proportion as the distance increases.

intercepted by *a b* at 1 foot. At 2 feet a considerable portion and at further distances increasing proportions of the external rays pass on without interruption, while the rays intercepted are the more central rays until at 15 feet they would be practically parallel.

Lenses.—Lenses of primary form are solid transparent media, such as glass or rock crystal, bounded by a polished spherical surface on one or both sides, and having the property of changing the course of rays of light falling on one of their surfaces in a direction parallel with the principal axis, so as to cause them either to *converge* to a given point—the principal focus—or to *diverge* as if they proceeded from the principal focus.

Cylindrical lenses will be described separately.

Convex and Concave Lenses.—The lenses which are chiefly used for optical and ophthalmoscopic purposes are centric convex and concave lenses—the former having two convex surfaces which are portions of equal spheres, double convex lenses, and the latter having two concave surfaces which are also portions of equal spheres, double concave lenses. They are sometimes designated bi-convex or equi-convex, and bi-concave or equi-concave lenses.

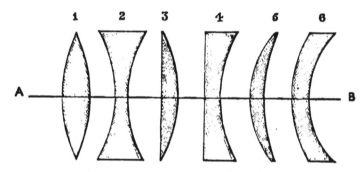

FIG. 18.—CONVEX AND CONCAVE LENSES.

1, bi-convex; 2, bi-concave; 3, p!ano-convex; 4, p'ano-coi cave; 5, meniscus · 6, concavo-convex; AB, line passing through the centres of curvature of the lenses, and which is perpendicular to the plane surfaces of 3 and 4.

The *convex* or *converging* lenses are thicker in the middle than at their edges; the *concave* or *diverging* lenses are thinner in the middle than at their edges. A double convex lens may be regarded as composed of two prisms with their bases joined at the centre; and a double concave lens of two prisms having their bases outwards, and their edges meeting at the centre. Rays of light whose passage has been limited to the lateral parts of a convex or concave lens are acted upon as they would have been if they had passed through prisms with their bases in similar directions. In either kind of lens, a line joining the centres of curvature of the two surfaces is the *principal axis* of the lens, any straight line other than the *principal axis* which passes through the centre of a lens is designated a *secondary axis*. The more convex the lens, or in other words, the shorter the radii of curvature of the two surfaces, the greater its converging power and the less the distance of its principal focus; the more concave the lens, or again the shorter the radii of curvature of its two surfaces, the greater its diverging power.

Other forms of lenses have one surface plane and the other convex or concave, and are called *plano-convex* and *plano-concave* lenses; or one surface may be convex and the other concave, and with such a lens if the convexity be in excess it is styled a *meniscus*

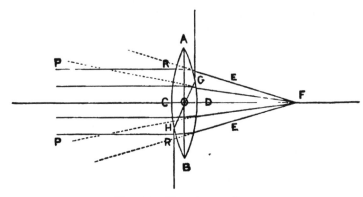

FIG. 19.—EQUI-CONVEX LENS.

A B, diameter of lens ; CD, principal axis ; G H, a secondary axis ; O, optical centre ; F. principal focus ; C, centre of curvature of ACB ; D, centre of curvature of ADB ; PR, PR, incident rays parallel with CD, converge as EE, after passing through the lens, and meet at F.

lens (like the crescent moon); if the concavity be in excess, a *concavo-convex* lens. The forms of these various descriptions of lenses are shown in fig. 18, 1 to 6.

The *principal focus* of a bi-convex lens will be at the point to

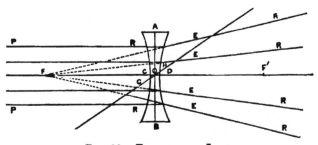

FIG. 20.—EQUI-CONCAVE LENS.

A B, diameter of lens ; CD, principal axis ; G H, a secondary axis ; PR, PR, incident rays parallel with CD, are rendered divergent (ER, RR) after passing through the lens, as if they had started from F, the principal focus ; O, optical centre of lens ; F', posterior focus.

which rays of light falling on one of its surfaces parallel with its principal axis, and undergoing refraction, firstly by passing through its substance, and secondly by leaving it for the rarer medium of the air, are caused to meet beyond its other surface. The principal focus of a bi-concave lens is, on the other hand, on the same side as the surface on which the incident rays of light fall (see figs.

19 and 20). The distance of this focus in both kinds of lenses is measured from the *optical centre* of the lens. The measurement was formerly made in inches and parts of inches ; but is now commonly made in metres or parts of a metre in optical measurements. The optical centre of a double convex or double concave lens, with surfaces of equal sphericity, is situated within the lens practically at the point where the principal axis of the lens is intersected by a diameter of the lens. It is at an equal distance from the two surfaces when the radii of their curvatures are equal. If the sphericity of the two surfaces is unequal, the distances of the optical centre from the two surfaces will not be alike, but will be in direct proportion to the radii of curvature of the two surfaces, and the focal distance will also vary in equal proportion according to the surface on which the incident rays fall. If the lens be plano-convex, the optical centre will be on the convex surface ; if plano-concave, on the concave surface. If the lens be a meniscus or a concavo-convex lens the optical centre will be outside the lens, and its distances from the two surfaces will still differ in proportion as their radii of curvature differ, while the distance of the principal focus will also proportionably differ according as the incident rays fall on one or other of its two surfaces. In common speech, for the sake of brevity, the principal focus of a convex lens is usually spoken of as *the focus* of the lens; and the distance of this focus is used to particularise the lens. Thus taking, *e.g.*, a lens whose principal focus is at 10 inches from its centre, it may be designated as a convex lens whose focus is at 10 inches, or a lens of 10-inch focus, or still more briefly as a 10-inch lens ; or, according to the metrical system, is a lens of four *dioptrics*, having its principal focus at a distance of one-fourth of a metre, or a 4 D lens. Convex lenses have what is called *real* foci (see p. 36).

If the rays falling on a double convex lens are divergent, and issue from a distance beyond that of its principal focus, they will be brought to a focus farther off than its principal focus ; if they diverge from a point at the distance of the principal focus, they will be refracted as parallel rays ; if they are convergent, they will be rendered more convergent and will be brought to a focus nearer to the lens than its principal focus. When rays diverging from a focal point fall on one surface of a convex lens and converge to a focus on the other side of the lens, and a line joining the two foci would pass through the centre of the lens, they are called *conjugate foci*. Such two foci are mutually so related that if the emitted rays proceeded from either of them the rays would meet in the other focus. It is on this principle that images of objects are formed ; every radiant point of the object emits a pencil of rays which is caused to converge to its conjugate focus in the image.

Rays falling on a double concave lens parallel with its principal axis are caused to diverge after passing through it to the same extent as they would diverge if they were proceeding from a point at the same distance from its centre as its principal focus is. In other words, the emergent rays of such a lens, if produced backwards, would meet in the principal focus of the lens on the same side as the incident rays. If the rays falling on the double concave lens are divergent instead of parallel rays, they will be rendered proportionably more divergent, just as if they came from a point

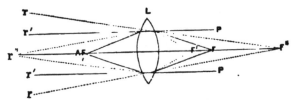

FIG. 21.—DOUBLE CONVEX LENS.

rr, convergent rays, brought to a focus at F', between the lens and the principal focus, F, to which *r's'*, parallel rays are brought; *r''*, diverging rays arising in front of AF, the anterior focus, meet at F'', beyond the principal focus (F). *r''* and F'' being on the principal axis are conjugate foci, as are also AF and F.

nearer to the lens than its principal focus. If they are convergent, and they converge in a direction toward any point between the lens and its principal focus, the refracted rays will be convergent also; if they converge to a point at the distance of the principal

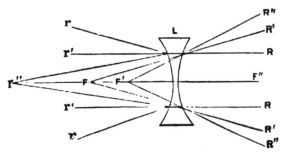

FIG. 22.—DOUBLE CONCAVE LENS.

r, *r'*, *r''*, incident rays; R, R', R'', emergent rays.
1. *r*, *r*, convergent rays, after passing through the concave lens, L, pass out as parallel rays, R, R.
2. *r'r'*, rays parallel with the principal axis emerge as diverging rays, R', R', with a direction as if they had started from F, the principal focus.
3. *r''*, *r''*, divergent rays become still more divergent, as though they had started from F', a point between the lens and the principal focus.
F, posterior virtual focus of the lens.

focus, they will be refracted as parallel rays; if they converge towards some point beyond the distance of the principal focus, the refracted rays will be proportionably divergent, just as if they were proceeding from some point further off from the lens than the distance of its principal focus. Concave lenses have what are called *virtual* foci (see p. 36).

Regarding the eye as a combination of lenses having together a certain converging power, when a convex lens is placed closely before it, the converging action of the eye on rays entering it from objects in front is necessarily added to ; when a concave lens is similarly placed before the eye, its converging action on these rays is lessened. The convex lens, therefore, is rightly designated in this respect a + lens ; the concave lens a — lens. It is convenient, too, in calculating the focal distances or powers of lenses required to correct abnormal conditions of the refractive media of the eye, to apply the sign + to converging lenses, and — to diverging lenses. Thus, a convex lens of 10-inch focus is designated a + 10-inch lens, or, in metrical numeration, a + 4 D (4 dioptric) lens ; while a concave lens of 10-inch focus is noted as a — 10-inch lens, or — 4 D lens.

When a convex lens is placed before the eye, it causes an advance of the nodal point of the eye, and thus increases the size of the retinal image; when a concave lens is similarly placed, it causes the nodal point to recede, and so lessens the size of the retinal image.

Numeration of Lenses.—Lenses have been hitherto numbered according to one or other of two systems ; one known as the duodecimal or *inch system*, the other as the *metric system*. Although the newer system, the metric, is certainly the more advantageous, and is doubtless destined to supersede all others for optical practice, it is not yet universally used by surgeons and opticians in this country, while the inch system of measurement still continues the one usually employed in commerce and is still maintained in the numeration of lenses made use of in various optical instruments, as microscopes, telescopes, and others. Until the metric system is universally adopted, the principles of both systems should be understood, and surgeons whose duties require them to engage in optical work should be familiar with the means of readily changing the measurements from one system into those of the other.

According to the inch system or older method of numeration, lenses are numbered according to the position of their principal focus. A lens whose principal focus is one inch from its centre is taken as the standard of unity. Starting from this high point, a series of lenses, say 25 + and 25 — lenses, descend in succession with lessened refracting power, and, consequently, with the position of the principal focus proportionally increased in distance.

In the metrical system, which was first proposed at Paris in 1867 but only brought into practical use in 1876, the lenses are not numbered according to the position of the principal focus, but according to their refractive power. A lens of such refracting power that its principal focus is at a distance of 1 metre, or 100 centimetres, is taken as the standard of unity. This in ordinary language is designated a *dioptric*. Starting from this low point, a series of lenses ascends with successively increased refracting power, and, consequently, with the position of the principal focus proportionally shortened in distance. Thirty lenses constitute the usual series in this system— that is, 30 + and 30 — lenses.

Power of a Lens in relation to the Position of its Principal Focus.—By the inch system of measurement, the strongest lens being taken as No. 1, the numbers attached to all other lenses represent fractional *parts*, as regards refracting power, of the standard of unity. The *power* of each lens is in an inverse ratio to the distance of its principal focus. Inverting the number expressing the principal focal distance of a lens, therefore, gives a ready means of expressing its power. Thus, taking a lens with power to cause parallel rays to converge to a focus at a distance of 1 inch from its centre as the standard of unity, another lens by which similar rays are brought to a focus at 2 inches from its centre, has manifestly only half the dioptric power of the former ; another, whose focus is at 3 inches, has only one-third of the converging power of the first ; at 10 inches, one-tenth ; and so on through the whole series. In the first instance the focus is 1 inch, and the power is expressed as 1 ; in the second, the focus is 2 inches (or $\frac{2}{1}$), and the power is $\frac{1}{2}$; in the third and fourth cases, the foci being 3 and 10 inches respectively (or $\frac{3}{1}$ and $\frac{10}{1}$) the powers are $\frac{1}{3}$ and $\frac{1}{10}$; and similarly through all the series.

By the metrical system of measurement, as a lens of very feeble power is taken as No. 1, the numbers of other lenses represent *multiples* of the standard of unity, and the *distance of the principal focus* is in an inverse ratio to the *power* of the lens. Inverting the power, therefore, gives the position of the focus. One dioptric, or a lens of one metre focus, being the standard of unity, a lens which has the power of two dioptrics, or a No. 2 lens, has its focus at a distance of half a metre, or 50 cm. ; another, which has the power of four dioptrics, or a No. 4 lens, has its focus at one-fourth of a metre, or 25 cm. ; of ten dioptrics, No. 10 lens, one-tenth of a metre, or 10 cm., and so through the whole series of lenses.

Comparison of the Two Systems of Numeration of Lenses.— The difference in the practical working of the two systems is very great. Two important changes have been effected by the introduction of the metric system. The potential intervals between the successive numbers of lenses have been simplified, and errors from the different lengths of inches in different countries, and consequent variations in lenses nominally of the same power, have been avoided.

In the inch system of numeration, as regards the difference in power between lenses of successive numbers, no two intervals are alike throughout the series. The difference in power between a 2-inch lens and a 3-inch lens, or lenses with powers equal to one-third and one-half of the standard of unity, is equal to a 6-inch lens, or lens of one-sixth power ; between a 3-inch lens and a

4-inch lens is a 12-inch lens, or lens of one-twelfth power; between a 4-inch and 5-inch lens is a 20-inch lens; and the values of the intervals are equally irregular throughout the whole series.

In the metrical system, the value of the interval between the adjoining whole numbers is always alike. The difference in power is uniformly that of one dioptric. A No. 2 lens has twice the power of the No. 1 lens; a No. 3 lens is one dioptric stronger than the No. 2 lens; a No. 4 lens is equally one dioptric stronger than the No. 3 lens. A No. 10 lens has the power of 10 dioptrics, and the same regularity prevails throughout. In working on this system it is always known that on taking a next higher whole number among the convex lenses, the converging power is increased by the strength of one dioptric, and in taking the next lower number it is decreased by the strength of one dioptric; or, in dealing similarly with concave lenses, that the diverging quality is increased or decreased to similar amounts. The relative strength to one another of all the lenses in the metrical series is thus at once made known by their numbers.

When the inch system is employed, the difference in value of the inches proper to different countries has to be taken into account. The English, French, Austrian, and Prussian inches all differ in length, and, though unimportant when dealing with weak lenses, these differences if ignored are liable to lead to important errors in optical practice when dealing with lenses of high powers. The Paris inch is ·02707 of a metre, the English inch ·02540; and thus the Paris inch exceeds the English inch by one-sixteenth, or, in other words, 16 Paris inches are equivalent to 17 English inches. The metre is of course the same in all places, and thus the difficulties arising from the different values of inches in different countries are avoided.

Another advantage of the metrical system of numeration of lenses is stated to be that the necessity no longer exists for using vulgar fractions, which is unavoidable in working with the inch system. If the most powerful lens of a series be taken as the standard of unity, all the other weaker lenses must necessarily be fractional parts; if the weakest lens of a series be taken as the standard, all the other stronger lenses will be multiples of it. To those who are not ready arithmeticians, calculations of multiples are easier than calculations of fractions. This, however, only applies to whole numbers. If parts of whole numbers are introduced into the series, fractions, either decimal or vulgar, must be used in calculations concerning them. In certain parts of the metrical system of lenses the intervals are fractional, and in using these parts decimal fractions are employed.

The Series of Metrical Lenses.—Although the number of metrical lenses in use, as before stated, is 30, the highest in the series is 20 D, having a focal length of one-twentieth of a metre, or 2 inches.

Under the inch system, although a 1-inch lens is taken as the standard of unity, the highest number employed in practice is also a 2-inch lens. In the metrical arrangement, the separation of a whole dioptric would have made too great a difference in the lenses of low power at the commencement of the scale. On this account the lenses have been added to by quarters up to 3 dioptrics, and by half dioptrics from 3 up to 6 dioptrics. From 6 up to 20 dioptrics, the differences are too slight to be broken. A table is given which shows the composition of the whole series of metrical lenses and their equivalents on the inch system. (See p. 28.)

To convert Lenses numbered on the Inch System into their Equivalents in the Metric System.—If the number of the lens be in Paris inches, invert it, and multiply by 36; if in English inches, multiply its inverted number by 40; the result will be the corresponding dioptric number. Thus a 12″ lens, Paris measurement, whose power is $\frac{1}{12}$″ multiplied by 36, is equivalent to a metrical lens of 3 D; or a 10″ lens, whose power is $\frac{1}{10}$″, English measurement, multiplied by 40, is equivalent to a lens of 4 D.

Or, more simply, if the lens be numbered in Paris inches, divide 36 by the number; if English inches, divide 40 by the number. Thus a 12″ lens in the former case is $= \frac{36}{12}$ or 3 D; the 10″ in the latter is $= \frac{40}{10}$ or 4 D; a 14″ lens, Paris inches $= \frac{36}{14} = 2.57$ D; a 16″ lens, English inches $= \frac{40}{16} = 2.50$ D.

To convert Lenses numbered on the Metrical System into their Equivalents in the Inch System.—If required to be in Paris inches, divide the metric number by 36; if in English inches, divide the number by 40. Thus a 4 D lens has for its equivalent in the inch system, Paris inches, $\frac{36}{4}$″ or $\frac{1}{9}$″, or a 9″ lens; in English inches, $\frac{40}{4}$″, or $\frac{1}{10}$″, or a 10″ lens.

To find the Measure of the Principal Focus of Lenses numbered on the Metric System.—If the distance be required in centimetres, divide 100 by the dioptric number; if in Paris inches, divide 36 by this number; if in British inches, divide 40 by the number. Thus, taking for example a 4 D lens, its focal distance is $\frac{100}{4}$, or 25 cm.; or $\frac{36}{4}$, or 9 Paris inches; or $\frac{40}{4}$, or 10 British inches. Or again, taking a 0.50 D lens, its focal distance is $\frac{100}{.50}$, or 200 cm.; or $\frac{36}{.50}$, or 72 Paris inches; or $\frac{40}{.50}$, or 80 English inches.

Aberration, Spherical.—Spherical lenses, in refracting homogeneous rays of light, are subject to what is termed *Spherical Aberration*, i.e. the rays refracted by different parts of the lens do not all intersect at precisely the same focal distance from the lens. The rays which pass through the lens near its circumference are refracted to a focus nearer to the lens than those which pass through its central portion, and this difference, of course, renders the image of an object less clear and defined. Spherical aberration is the more marked in degree in proportion as the *aperture of the lens*, or the angle obtained by joining the edges of the lens to the site of its principal focus, is increased; in lenses whose angular apertures do not exceed about 10°, the refracted rays practically meet at one and the same focus.

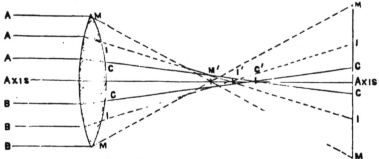

FIG. 23.—SPHERICAL ABERRATION.

AAA, BBB, incident parallel rays of light; MM, marginal; II, intermediate; CC, more central rays; M', focus of the marginal rays; I', focus of intermediate rays; C', decussation of the more central rays corresponding with the principal focus of the lens.

Aberration, Chromatic.—Another kind of aberration to which lenses are liable results from the different degrees of refrangibility of the simple colours, and is termed *Chromatic Aberration*. Since a convex lens consists, as it were, of a series of narrow prisms united at their bases, rays of light, when they pass through it, are not only refracted, but the light to a certain extent is decomposed. When this decomposition occurs in a noticeable degree, the violet rays, being most refrangible, intersect at a point nearer to the lens

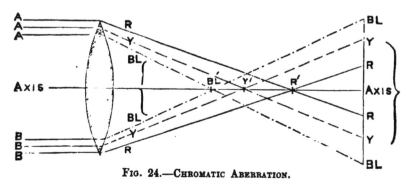

FIG. 24.—CHROMATIC ABERRATION.

AAA, BBB, incident parallel rays of homogeneous colourless light emerging as coloured rays; R, red; Y, yellow; Bl, blue; R', Y', BL', decussating points of the corresponding rays. It is thus apparent that Spherical and Chromatic Aberration are quite the converse of each other. The marginal rays cutting the axis first (M') in the former case, and last of all in the latter (R'). The more central, the intermediate, and the marginal rays in the former case hold the same relative position after decussation, while the order of the colours is inverted in the latter. The brackets show the area of white light under ordinary circumstances, both in front and behind the focal distance.

than all the other coloured rays; while the red rays, being the least refrangible, form their focus at a point on the axis of the lens farthest from the lens. This quality interferes with the perfect definition of images, and renders them defective by imparting colours to their edges. Practically, however, as in the case of an ordinary lens, red and yellow are only to be seen anterior to the focal point, while blue alone is visible after the decussation of the rays. This is shown by the position of the brackets in fig. 24.

Table of Metrical Lenses and their Relative Values in English Inches.—The French metre being equivalent to 39·3707 English inches, as before shown, for practical purposes this may be regarded as 40 English inches. The calculations in the fourth column of the following table have been made on that basis.

Dioptric number.	Focal length in metres.	Focal length in centimetres.	Approximate focal length in English inches.	Dioptric number.	Focal length in metres.	Focal length in centimetres.	Approximate focal length in English inches.
0·25	4	400	160	5	0·200	20	8
0·50	2	200	80	5·50	0·182	18·2	7·33
0·75	1·333	133	53	6	0·166	16·6	6·66
1	1	100	40	7	0·143	14·3	6
1·25	0·800	80	32	8	0·125	12·5	5
1·50	0·666	66·6	26·66	9	0·111	11·1	4·50
1·75	0·571	57·1	23	10	0·100	10	4
2	0·500	50	20	11	0·091	9·1	3·75
2·25	0·444	44·4	18	12	0·083	8·3	3·33
2·50	0·400	40	16	13	0·077	7·6	3
2·75	0·363	36·3	14·54	14	0·071	7·1	2·85
3	0·333	33·3	13·33	15	0·067	6·6	2·66
3·50	0·286	28·5	11·50	16	0·062	6·2	2·50
4	0·250	25	10	18	0·055	5·5	2·25
4·50	0·222	22·2	9	20	0·050	5	2

Cylindrical Lenses.—A cylindrical lens is one which has the curvature of a cylinder instead of that of a sphere ; it is, in fact, a portion of a segment of a cylinder. The opposite sides of a cylindrical lens are parallel with each other in the direction of the axis of the cylinder of which it is a segment, while there is more or less curvature between them in all other directions. The curvature is greatest in the direction perpendicular to that of the axis of the cylinder. The foci of cylindrical lenses may be either positive or negative. There are several kinds of cylindrical lenses.

Simple Cylindrical Lenses.—Ordinary cylindrical lenses are cylindrical on one side only and flat on the other. They are known as *plano-convex* or *plano-concave* cylindrical lenses. But, like spherical lenses, they may be bi-convex or bi-concave, or convex-concave or concavo-convex lenses. In these, which have both surfaces cylindrical, the axes must be parallel. The plano-convex and plano-concave cylindrical lenses are the lenses chiefly employed for trying and correcting simple forms of astigmatism.

The illustrations which follow are explanatory of the points mentioned in this and in the preceding paragraph. Fig. 25

exhibits the principal features of a cylinder so far as concerns its optical relations to cylindrical lenses; figs. 26 and 27 indicate the

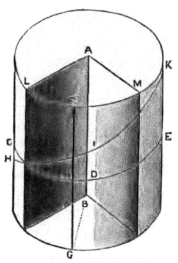

FIG. 25.—PORTION OF CYLINDER.

LAM, sector of the cylinder; AB, axis of the cylinder; LA, MA, two radii of cylinder; LFM, arc of the sector; CDK, principal curvature perpendicular to axis; FG, line parallel with axis; KIH, a secondary curve of the cylinder.

mode of construction of a plano-convex and plano-concave cylindrical lens, and show the aspects of their two principal sections.

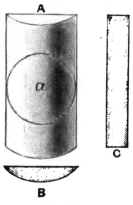

FIG. 26.

A, plano-convex segment of a cylinder; a, cutting of a plano-convex cylindrical lens; B, transverse section; C, longitudinal section through the axis.

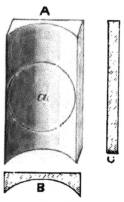

FIG. 27.

A, plano-concave segment of a cylinder; a, cutting of a plano concave cylindrical lens; B, transverse section; C, longitudinal section through the axis.

Bi-cylindrical Lenses.—These are lenses with both surfaces cylindrical, but with the axes of the cylinders perpendicular to each other. Their construction is shown in figs. 28 and 29.

Spherico-cylindrical Lenses.—These are compound lenses, of which one surface has a spherical, the other a cylindrical curvature. They are chiefly used in the correction of compound and mixed forms of astigmatism. Their construction is shown in figs. 30 and 31.

Action of Cylindrical Lenses upon Light.—Rays of light falling on a simple cylindrical lens are acted upon differently according as they

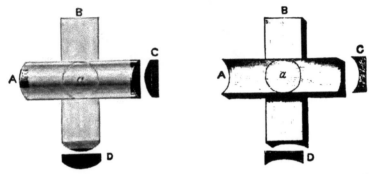

FIG. 28.—CONVEX BI-CYLINDRICAL LENS.   FIG. 29.—CONCAVE BI-CYLINDRICAL LENS.

A, anterior segment; B, posterior segment; C, vertical section; D, transverse section.

fall upon the lens in a plane coincident with the axis of the cylinder of which it is a segment, or in a cross direction. Rays incident in the plane of the axis of the cylinder pass through it as through glass with opposite parallel surfaces, or as if it were a piece of plate glass, and form on a screen a line of light. Rays incident in a plane at right angles to the axis are acted upon according to the amount of convexity or concavity of the glass in the direction

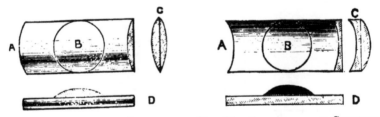

FIG. 30.—SPHERO-CONVEX CYLINDRICAL   FIG. 31.—SPHERO-CONCAVE CYLINDRICAL .
LENS.   LENS.

mentioned, and converge to a point or diverge as if issuing from a point. The more convex or the more concave the surface, the more convergent or more divergent the rays will be rendered after their passage through the cylinder in this direction, just as happens with spherical lenses. The essential difference between a spherical lens and a simple cylindrical lens is that in the spherical lens all the rays which fall on it are altered in direction by their passage through it,

while in the cylindrical lens some of the incident rays pass through
it unaltered in direction, and the remainder of the rays are more or
less altered in direction. A spherical lens brings the rays proceeding from a luminous point to a focus in the form of a point,

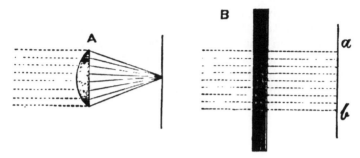

FIGS. 32 AND 33.—ACTION OF SIMPLE CYLINDRICAL LENSES ON LIGHT.

A, transverse section with convergent rays.   B, axial section with parallel rays.
Combination of A and B resulting in a linear focus, *ab*.

while a cylindrical lens brings the rays from a similar point of
light to a focus in the form of a line instead of a point.

Prisms.—Prisms of different powers are used in certain parts of
ophthalmic practice, especially in the investigation and treatment of
strabismus and *diplopia.*   Prisms for optical
purposes essentially consist of a solid piece of
crown glass contained between five plane surfaces, of which three—two *sides* and a *base*
—are rectangular, and two—the *ends*—triangular, these latter being at right angles with
the base of the prism, and therefore parallel
with each other.   The line of junction of the
two inclined surfaces, or sides, is the *edge*
of the prism ; each triangular end, if perpendicular to the base, represents a *principal
section* of the prism (see fig. 34).   In order to
adapt them for use in front of the eye, however,
the angles are sometimes rounded off, and
a shape thus given to them which is nearly
circular.

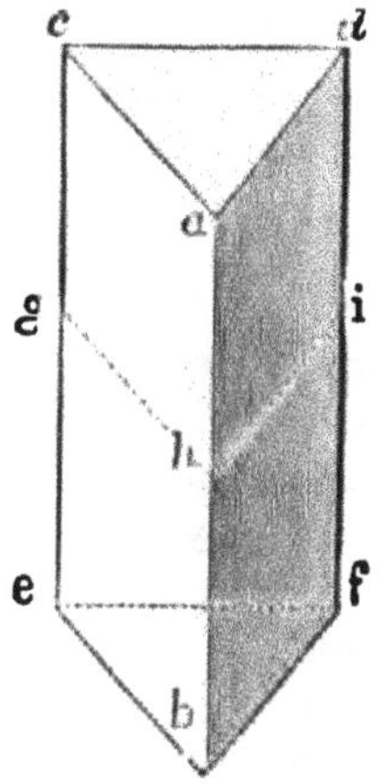

FIG. 34.—THE PRISM.

cabe, dabf, two rect-
angular sides ; *ccjd,*
base ; *cad, ebf,* two
ends, or principal
sections ; *ab,* edge of
prism; *ght,* refracting
angle of prism.

When rays of light are made to fall upon
one of the sides of a prism, and to pass through
it, they are refracted towards its base, and the
degree of their deflection on issuing from the
other side of the prism varies according to the
size of the angle inclosed between the sides through which the refracted light has passed, or, in other words, according to the *angle
of refraction* of the prism.   As a consequence of these qualities,

when an object is looked at through a prism, its apparent position is changed; it seems to shift its direction towards the edge of the prism, and the distance to which it appears to be shifted from its normal position increases with the increase in the refracting angle of the prism (see fig. 35). Taking advantage of these facts, the rays from a given object may be made to impinge upon any part of either retina, by the use of prisms of different degrees of strength placed in suitable positions before the eyes; and by these means the presence or absence of binocular vision may be determined in doubtful cases, diplopia may be counteracted, and single vision restored, or the relative strength of the several motor muscles of the eyes may be tested.

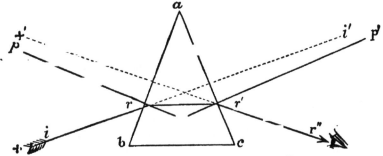

FIG. 35.—REFRACTING PROPERTIES OF THE PRISM.

+, object; ir, ray in air incident on ab, one of the sides of the glass prism, abc; pr, perpendicular to ab; ii', direction of incident ray if no prism intervened; rr', incident ray refracted in prism toward perpendicular pr; $p'r'$, perpendicular to side ac of prism; $r'r''$, emergent ray from glass prism into air, refracted from the perpendicular, $p'r'$; +', direction in which the object + is seen by an eye at r''.

The effect of placing a prism before a squinting eye may be seen in fig. 36. In diagram A, a right eye with fixed internal squint sees the image of an object displaced outwards; but with a prism of suitable refracting power placed before it with its base outwards, as shown in diagram B, is caused to see the image in its true direction.

When a pair of prisms of equal refracting angles are placed before the two eyes with their bases outwards, and an object situated at a limited distance is looked at, it appears at first to each eye to be displaced in a direction towards the edge of the prism or inwards, and the strain on the muscles of convergence of the two eyes, the recti interni, in order to preserve single vision is increased; when the prisms are placed with the bases inwards the object appears to each eye to be displaced outwards in the direction of the edges of the prisms, and not only is the extra strain on the recti interni muscles then taken off, but the need for the normal action of these muscles may be lessened.

The strength of the convergent power of the eyes may be measured by causing the two eyes to look in the manner just described at an object situated in a plane midway between them, through two prisms with their bases directed outwards. The object which appears to be displaced in the direction of the edges of the two prisms respectively, or across the median line, if the prisms be of sufficient strength, will be seen double, and, if the power of the prisms is greater than that of the internal recti muscles to fuse the double images into a single one, the two images will remain at a certain distance apart. The highest refracting angle in the prisms which can be overcome by the muscular efforts, as shown by the fusion of the two images into a single image, affords the measure of the extent to which the converging faculty can be exerted.

In like manner, if the bases of the prisms be held inwards, the

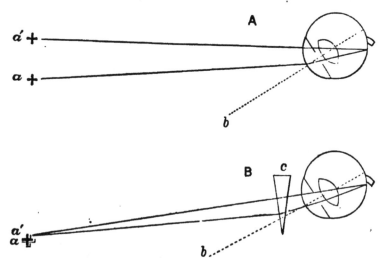

Fig. 36.—Internal Strabismus (Right Eye).

A. Without prism. a, object; a', position of image, homonymous, or with external displacement; b, optic axis.
B. With prism. a, object, with which a', the image, is rendered coincident by the action of the prism, c.

limit in diverging power of the external recti muscles may be measured.

The prisms supplied in the cases of eye lenses and prisms in ordinary use have angles of refraction varying from 3° to 24°.

Trial Case of Lenses.—A case of lenses of various descriptions and powers employed for purposes of optical investigation. A complete trial case usually includes full sets of the convex, concave, and cylindrical lenses just described, each bearing a distinguishing mark of its quality and power, together with a series of prisms, coloured glasses, stenopœic diaphragms, opaque discs, and two pairs of lens-holders or trial frames.

D

Trial Frames or Lens-holders.—Two kinds of frames are required for holding lenses to serve as spectacles, and are generally
supplied with each trial case. One with a single groove is arranged
for holding a pair of spherical lenses; the other with a double
groove for holding cylindrical lenses in addition to the spherical
lenses. The posterior grooves are intended to hold the spherical
lenses; the anterior, the cylindrical lenses. The metal rim of each
of the hemispheres within which the cylindrical lens is placed is
divided into degrees, plainly marked on the surface; so that any
required inclination may easily be given to the axis of the cylindrical lens held by it.

**Ready Tests of Convex, Concave, and Simple Cylindrical Spec-
tacle Lenses.**—It is frequently necessary to ascertain rapidly the
nature of a lens. This can be readily done in the following

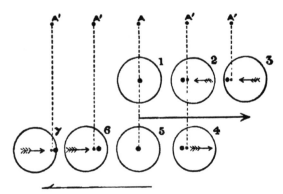

FIG. 37.—CONTRARY MOVEMENTS OF CONVEX LENS AND IMAGE.

A, a black dot, supposed to lie in the axis of 1, a + lens; A′A′A′A′, the same dot holding
intermediate and marginal positions according as the lens is moved to the right or left.
The larger arrows show the movement of the lens, and the smaller ones the movement
of the enlarged image, which always reaches the margin of the lens in advance of the
object.

manner :—On holding the lens a few inches in front of some
printed type and looking through it at the print while the lens is
moved in a plane parallel with the plane of the print, if the print
is enlarged and appears to move in an opposite direction to the movement of the lens, the lens is convex; if the print is diminished in
size and appears to move in the same direction as the lens, the lens
is concave; and if in either case the same occur equally in whatever direction the lens is moved, the lens is centric. If the respective movements just described only occur in one direction, the
print appearing to remain stationary when the lens is moved in the
opposite direction, the lens respectively is a simple + or — cylindrical lens.

Similar indications will be obtained if the lens be moved an inch
or so in front of the eye, and an object at a distance of several feet off
be looked at through it. If the lens be a centric convex lens, the

object will appear in all directions to move the opposite way to
the movement of the lens; if it be a concave lens, it will appear to
move in the same direction. If it be a simple convex cylindrical
lens, the dimensions of the object will be elongated in one direction,
and in this direction will appear to move opposite to the movement
of the lens; while it will seem to be shortened, but show no movement, when the lens is moved in a contrary direction. This latter
will be the line of direction of the axis of the lens. If it be a simple
concave cylindrical lens, the dimensions of the object looked at
through it will be lessened in one direction, and in this direction
will appear to move in concert with the movement of the lens;
while it will seem to be elongated, and show no movement, when
the lens is moved in the contrary direction. The last-named will
be the line of direction of the axis of the lens.

**To ascertain the Power of a Convex, Concave, or Cylindrical
Lens.**—This may be readily done in the case of a centric convex
lens, even without a trial case of lenses, by ascertaining the distance
of its principal focus. The rays of light from a window frame fifteen
or twenty feet away, or, as is often done in opticians' shops, from
the large letters on a shop front on the opposite side of a street, are
allowed to pass through the lens and are received on a piece of white
card or other suitable screen. At one distance only can a clearly
defined and sharp picture occur of the object from which the rays
have come, and the measure of this distance from the lens gives the
distance of its principal focus, and therefore of the power of the
lens under examination.

This cannot of course be done in the case of a diverging lens.
When the power of a centric concave lens is sought, it must be
ascertained by neutralising it by a convex lens of known power.
This is quickly managed with a trial case of lenses at hand. A
convex lens is taken from the case, placed in contact with the concave lens, and an object, such as the bar of a window frame, looked
at through the two lenses combined while they are moved in front
of the eye. If the bar appear to move in the opposite direction to
that of the lenses, the convex lens is stronger than the concave lens;
if the bar appear to move in the same direction, the convex lens is
less strong than the concave lens; when the bar, seen through the
two lenses, shows no movement on the lenses being moved, the
concave lens is exactly neutralised, and the power of the convex
lens being known, that of the concave lens is also known.

In the same way the power of a convex lens may be quickly
ascertained, if a trial case of lenses is at hand, by finding the
concave lens which exactly neutralises it.

In the case of a convex cylindrical lens, the principal focus, and
therefore its power, may be ascertained by measuring the distance
from the lens at which a point of light, as a candle placed at the
distance of 10 or 15 feet, is brought to a sharply defined line of

light on a screen. The power of a concave cylindrical lens must be found by neutralising it by convex cylindrical lenses of known power; and, indeed, both forms of cylindrical lenses can have their powers most quickly ascertained by neutralisation by opposite kinds of cylindrical lenses taken from a trial case of lenses, in the same way as concave and convex spherical lenses. The axes of the two opposite kinds of cylindrical lenses must be so placed as to coincide exactly in direction.

Army Optical and Ophthalmoscopic Case.—In the combined optical and ophthalmoscopic case which used to be supplied to officers of the Army Medical Department, in addition to the convex and concave lenses belonging to the ophthalmoscope, there was a pair of spectacles fitted with 10-inch convex or +4D lenses. Parallel rays falling on such lenses are caused to converge to foci at a distance of 10 inches from their respective centres; while divergent rays proceeding from points at 10 inches' distance from their centres are refracted by them as parallel rays.

Formerly spectacles with concave −6″ lenses were also supplied in this optical case; but the supply was discontinued because it was found that their employment could be dispensed with without inconvenience.

Images formed by Convex and Concave Lenses.—Images are representations of objects formed by concourse of the pencils of rays emanating from all the points of the objects which they represent. All the points of a perfect image are the conjugate foci of the rays which have proceeded from corresponding points of the object (see ' Conjugate Foci,' p. 21).

When an image is formed in the focus of a lens, and can be received on a screen, as at the focus of a convex lens for example, it is called *real* or *positive*; when it is not formed by the actual union of rays in a focus, but only appears to be so, and cannot be received on a screen, it is called a *virtual* image.

An image is *erect* when the object and image lie on the same side of the centre of the lens; is *inverted* when the object and image lie on opposite sides of the centre. The retinal image of any object situated in front of the eye is an example of a *real* and *inverted* image.

The diameters of the object and its image are directly as their distances from the centre of the lens; as they separate from this point, the farther off either is, the greater its proportionate size.

When an object is placed between a *convex* lens and its principal focus, an eye on the other side of the lens sees a *virtual* image of the object, erect, magnified, on the *same side* of the lens as, but at a greater distance from it than, the object. It is on this principle that microscopes are formed. If the object be farther off than the principal focus, but its distance be less than twice the focal length of the lens, the image is *real, inverted, magnified,* and on the *oppo-*

site side of the lens to the object; if the distance of the object be greater than twice the focal length of the lens, the image is *real, inverted, diminished,* and on the *opposite side* to the object.

When an object is placed in front of a *concave* lens, an eye on the other side of the lens sees an image of the object which is *virtual, erect, diminished,* on the *same side* of the lens as, and nearer to it than, the object. The image is diminished when the distance between the lens and the object is increased; but when the distance

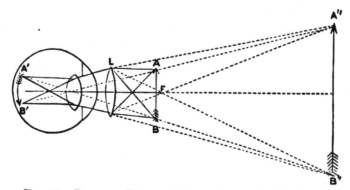

FIG. 38.—REAL AND VIRTUAL IMAGES (DOUBLE CONVEX LENS).

AB, object; A'B', real image on retina; A"B", virtual image; F, anterior focus; L, lens.
The lines of reference from the retina are strongly dotted.

of the object is a large multiple of the focal length of the lens, further increase of its distance does not appreciably alter the distance of the image, or, consequently, its size.

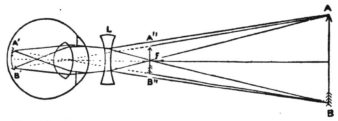

FIG. 39.—REAL AND VIRTUAL IMAGES (DOUBLE CONCAVE LENS).

AB, object; A'B' real image on retina; A"B", virtual image; F, anterior focus; L, lens.
The lines of reference from the retina are strongly dotted.

Composition of Optical Lenses.—Lenses for spectacles are either made of crown glass or colourless quartz, and it is useful for surgeons to have some knowledge of their respective qualities and occasional defects. Lenses made of quartz are described by the opticians who sell spectacles as 'pebbles,' or 'rock crystal,' and are usually recommended by them as being clearer and cooler to the eyes than glass. The clearness of view through pebbles depends on the quality and manipulation of the quartz, and can very rarely, if ever, be equal to that of the best crown glass; any superiority in coolness as regards the eyes is very problematical. Glass is a bad conductor of heat, and quartz conducts it better than glass; hence a quartz lens feels relatively cool when applied to the tongue, as it abstracts the heat from it more quickly than a glass lens. This quality is sometimes turned to account as a rough-and-ready test for distinguishing between a rock crystal and glass spectacle-lens. The only real advantage of 'pebbles' is their greater hardness, so that their surfaces are not scratched and dulled so easily as those of glass lenses. Unless the quartz is unusually pure and transparent, is scientifically cut and shaped as regards refraction, it is a decidedly inferior material to good crown glass for optical purposes; and it is doubtful whether pebbles do not remain more trying to the eyes, even when proper attention is given to the requisite qualities just named, than spectacles of the best crown glass, such as is specially manufactured for spectacle lenses. Quartz has the quality of double refraction, and although its effects may not be perceived by the eyes in thin lenses of low powers, the images of objects on the retina must be in some degree less perfect in distinctness in consequence, the more so if the quartz is not well cut, and a certain amount of accommodation will be exerted in trying to render them sharp and single. The only way to cut rock crystal so as to avoid double refraction is to cut it exactly perpendicular to the axis of the crystal, so that the axis of the lens which is formed from it may coincide, or be parallel with the axis of the original crystal. The operation is difficult except in comparatively rare specimens of perfectly crystallised quartz, requires special care, and renders the lenses more costly.

It is sometimes desirable to establish whether a pebble lens is properly cut or not. This can readily be done by looking through the lens placed in a tourmaline forceps, which may be procured at the establishments of all good opticians (see fig. 40). The blades of the forceps consist of two thin plates of tourmaline, one of which can be revolved in front of the other, and they admit of a lens being placed between them. On turning the movable tourmaline plate, when it arrives in a certain position the transmitted light becomes polarised, and is as it were extinguished. If a quartz lens is now placed between the tourmaline plates, the light is depolarised, and the field becomes luminous; if the lens were a glass lens, the field would remain dark. This serves to distinguish between a glass and a pebble lens. On further examination of the quartz lens, if it has been cut perpendicularly to the axis of the crystal, its middle portion will be seen to be occupied by a series of concentric coloured rings; if no such rings appear, it has been cut in a contrary direction. If the rings appear towards the edge of the lens, or are elliptical in form, it is defective, for it has been cut obliquely in respect to the axis of the crystal.

Glass lenses for spectacles should be made of pure, colourless, and perfectly homogeneous crown glass. They should approach as nearly as possible in limpid transparency to that of the dioptric media of a young and perfect human eye. Inferior lenses are made of a bad quality of crown glass, having more or less of a greenish tint, and not unfrequently contain microscopic air globules, specks, or

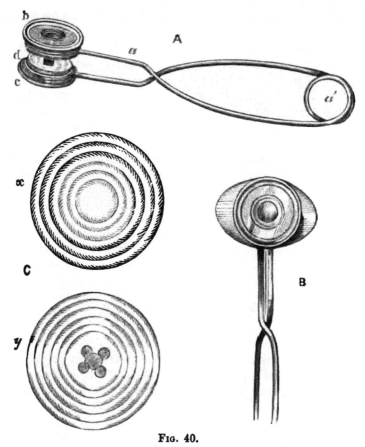

FIG. 40.

A, tourmaline forceps, half natural size, with the blades slightly open; *a*, simple wire frame-work, with loops to hold the brass mounting of the tourmalines (*b* and *c*), and a spring (*a′*) at the other end; *d*, cork lining, with a square central opening; B, a spectacle lens held between the blades of the forceps; c, pebble lenses as seen between the tourmalines (*x*), in the position to intercept light (axes crossed at right angles), and (*y*) when they transmit it (the axes being parallel).

striæ, owing to an imperfect mixture of the vitrifiable ingredients which enter into the composition of the glass. Such lenses interfere with a clear and even view of objects, they are unstable with respect to their transparency, are more or less unequal in their refractive power, and in consequence of these defects are proportionally trying and injurious to the eyes that use them.

Unless glass lenses are duly protected their surfaces are apt to lose their polish and transparency from being roughened by the fine gritty dust which collects on all objects exposed to the air. This is especially objectionable in a convex lens, for by such means it is rendered most dim at the centre on which it rests, the part where it is essential to retain perfect transparency. The superior hardness of rock crystal lenses gives them an advantage over glass under similar conditions of exposure.

Thickness and Weight of Glass and Pebble Lenses for Spectacles. It is important for the ease and comfort of those who have to wear spectacles that the lenses should not be needlessly thick and heavy. Although the refractive index of rock crystal is rather higher than that of crown glass, I have ascertained from Messrs. Pickard & Curry, ophthalmic opticians, that in the manufacture of lenses for spectacle purposes there is no difference in respect to thickness of substance or weight between glass and rock crystal lenses of corresponding powers. The following measurements show the average thickness and weights of spectacle lenses of a few given powers, as manufactured by competent opticians. Spherical bi-concave lens, -2 D, least thickness at centre, 1 mm.; thickness at circumference $1 \cdot 5$ mm.; weight of pair of lenses in frames 6 dwt., the lenses forming half the weight, the frames the other half. Spherical bi-convex lens, $+2$ D, thickness at centre 2 mm.; weight of pair of lenses in frames $5\frac{1}{2}$ dwt. Concave spherico-cylindrical lens, sph. -2 D $\bigcirc$ cyl. -2 D; thickness at centre in axial line 1 mm.; greatest thickness of circumference $1\frac{3}{4}$ mm.; weight of pair of lenses in frames $6\frac{1}{4}$ dwt. Convex spherico-cylindrical lens, sph. $+2$ D $\bigcirc$ cyl. $+2$ D, thickness through centre of spherical surface 2 mm.; weight of pair of lenses in frames, 6 dwt. Meniscus lenses and concavo-convex lenses are approximately the same in weight as bi-convex and bi-concave lenses of corresponding powers.

Arrangement of Lenses in Spectacles.—When lenses are applied in the form of spectacles for the purpose of relieving visual defects, several points of optical importance require attention. Glasses of the proper description and power may be ordered by a surgeon, but, however good in quality the lenses may be, unless they are rightly disposed by the spectacle maker their purpose may not only not be attained, but the eyes concerned may be much inconvenienced by them. Spectacles often require the inspection of the surgeon who has ordered them, to ascertain whether his directions have been carried out.

Spectacles should be so arranged that the visual axes pass through the centres of the lenses. For this purpose, in the first place, the lenses should be properly centred, or, in other words, so adjusted within the rims in which they are held, that the centres of the rims, and the centres of the lenses, are in exact correspondence. Although the lens may have been correctly ground in respect to the coincidence of its axis and the centres of curvatures, it is evident

that if it be not cut or shaped in perfect accordance with the shape
of the frame which receives it, the axis will be displaced, and the
vision of the wearer will be liable to be proportionally incommoded.

In the second place the frames carrying the lenses must be
accurately measured and adjusted to the eyes before which they
are placed, neither too narrow nor too wide. If the visual axes do
not pass through the centres of the lenses, the line of sight of each
eye will be either to the outer or to the inner side of the centre of
the corresponding lens. If the spectacles are convex, and the line
of sight is allowed to pass through the outer part of each lens, as
will be the case in too narrow spectacles, the rays arriving at each
eye from the object looked at are acted upon as if they had passed
through a prism with its base turned inwards ; if the line of sight
of each eye pass through the inner part of a convex lens, as will
happen when the lenses of the spectacles are placed too far apart,
the rays of light from the object looked at are changed in direction,
as they would be in passing through a prism with its base directed
outwards. The position of the two eyes must change in order to
meet the altered directions of the rays falling on the macula lutea.
In the former case the eyes will have to converge less, in the latter
case more ; and where the spectacles are used for near objects, as
in presbyopia for reading, the increased demand made on the action
of the muscles of convergence, and the disparity caused between
the exercise of convergence and that of the accommodation for the
distance at which the print is placed, will entail a sense of strain
and uneasiness (asthenopia) about the eyes. When concave spectacles are worn, and are not properly centred, the effects will be
the reverse of those just named for convex glasses. The necessity
for the visual axis passing through the centre of each lens involves
a slight difference in position of the lens, or in the adjustment of
the frame, according to the purpose for which the spectacles are
worn. If they are used for correcting distant vision, as in myopia
and hypermetropia, the two lenses must be parallel with one
another, and their two centres should be directly in front of the
pupils of the two eyes ; if for correcting vision of near objects,
they should be inclined at such an angle toward each other as to
allow the visual lines to pass through their centres to the visual
distance of the objects looked at through them. So with bi-focal
spectacles, in which the upper half of the spectacle is used for
distant objects and the lower lens for objects near at hand, the
upper lens sections should be placed parallel with the vertical
planes of the two eyes when looking forward, and the lower lens
sections connected with them at such an angle, about 75°, as will
preserve the same correspondence in direction when the eyes are
turned downwards towards objects close at hand, as in reading, the
centres at the same time being moved slightly inwards, to allow for
the convergence of the two eyes. When one side of a spectacle is

allowed to drop to a lower level than the opposite side, an occurrence which may not unfrequently be noticed, the visual axis of one or other eye no longer accords with the centre of the lens before it, and the discord leads to visual disturbance. A horizontal line joining the two centres of the lenses of a pair of spectacles ought to be exactly parallel with a similar line joining the centres of the pupils of the two eyes. A want of congruence in this respect is a frequent source of error when folding glasses are used.

Distances of Spectacles from the Eyes.—When spectacles are worn to correct refractive defects of the eyes, they should be so adjusted that the lenses are placed as near as possible to the anterior focus of each eye, *i.e.* about half an inch (13 mm.) in front of the cornea. When they are worn to correct failure of accommodation (presbyopia), the positions of the lenses are less important; if they are removed an inch or so further from the eye than its anterior focus, the only effect will be to modify slightly the apparent size of the objects looked at, and to alter the position of the near point of distinct vision. This is met by suitably adjusting the distance at which the objects looked at are placed.

Eyeglasses.—The use of a single eyeglass by any one who has the use of both eyes is not to be recommended, excepting in rare cases, when its purpose is to adjust the refractive power of the eye before which it is placed, so as to make it agree with that of the eye which is left free. If the effect of the glass is simply to assist the vision of the eye to which it is applied, while that of the other eye is left unattended to, the action of an eyeglass becomes more or less deleterious. The two eyes should always, as far as practicable, be exerted in concert; if the retinal image in one be habitually suppressed, the unused eye will in time become amblyopic. Eyeglasses are not allowed to be used at target practice at the School of Musketry, though spectacles may be worn.

Spectacle Frames.—The frames should be well fitted to the form of the face, should not press on the temples, but should only rest on the bridge of the nose and the tops of the ears, and, whatever the metal of which they are made, should be sufficiently firm to maintain the right positions of the lenses. When they are used for vision at near objects, or for work in which the face is required to be bent forward, it is advantageous for the extremities of their sides to be curved so as to hook behind the ears, and be thus prevented from shifting or dropping off.

Pantoscopic Spectacles.—A name given by opticians to spectacles in which a portion of the upper margin of each lens is cut away, so that they are nearly flat at the top and oval below. They enable the spectator to see over the lenses when he is not looking at near objects. The lenses, and the parts of the frames holding them, are set at an angle with the sides of the frames, so that the upper borders of the lenses are tilted to a certain extent forwards. They are suitable for emmetropic persons who have become presbyopic.

Some convex lenses are made perfectly flat above with the usual oval below ; while other concave lenses have this shape reversed, and are made oval above and flat below for myopic persons. As spectacles of these last two kinds are occasionally used by artists, they are usually sold under the name of 'sketching spectacles.'

Equi-convex and Equi-concave Spherical Lenses in Spectacles.— Rays which pass through the secondary axes of a spherical lens, although impinging obliquely, are not much altered in direction on leaving the lens, because the opposite surfaces near the centre are nearly parallel ; but as rays approach the circumference of the lens, they become more and more oblique relatively to its curvature, are more strongly converged on emergence, and so intersect at points nearer to the lens than its principal focus. This deviation, known as *spherical aberration*, has been explained elsewhere (see p. 27). It gives rise to a certain amount of imperfection in the use of bi-convex and bi-concave spherical lenses as spectacles when the eyes are so turned as to look obliquely through them. The spherical aberration is greater in a plano-convex or plano-concave lens, than it is in a bi-convex or bi-concave lens of corresponding power. As the rays most free from deviation are those which most nearly coincide with the principal axis of the lens, it follows that, as before mentioned, vision through a spectacle is best when the eye looks direct through its centre ; and it is for the purpose of maintaining vision in this direction that persons wearing spectacles, even when perfectly fitted, or using a glass, are in the habit of turning the head altogether towards an object in cases where persons without spectacles would merely turn their eyes.

Periscopic Glasses.—Lenses concave on one side, convex on the other. When they are sectionally crescentic in shape, *i.e.* when the convexity is in excess of the concavity, and the glasses act as convex glasses, they are designated *meniscus* glasses (μηνίσκος, a little moon, a crescent). When the concavity is in excess of the convexity and they act as concave lenses, they are styled *concavo-convex* glasses. In the former, the two surfaces meet if they are continued ; in the latter, under the same conditions, they do not meet (see fig. 41). They are positive or minus lenses, according as their convexity or concavity is respectively in excess. The distinctness of objects seen through a meniscus or concavo-convex lens is less interfered with by spherical aberration than when they are seen through bi-convex or bi-concave lenses of corresponding powers, and they further have the advantage of closer accordance in their form of curvature with the curvature of the front of the eye-ball, and in this way they enable the wearer to see more obliquely and to have a wider field of view, or rather to look round with less inconvenience when the eye only is turned without turning the head. Hence the name of *periscopic* glasses, given to them by Dr. Wollaston who advocated their employment for spectacles in preference to bi-convex and bi-concave glasses.

For a long time special difficulties were encountered in making satisfactory periscopic glasses at a moderate price, but they are now manufactured on the continent in large numbers of excellent quality, and may be obtained in England at as cheap a rate as

Fig. 41.

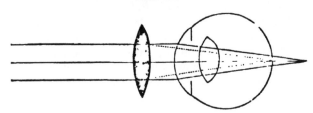

A. Hc. Eye corrected by a Bi-convex Lens.

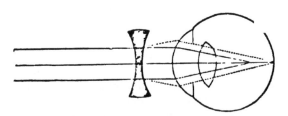

B. Hc. Eye corrected by a Meniscus Lens.

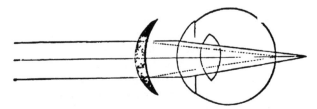

C. Mc. Eye corrected by a Bi-concave Lens.

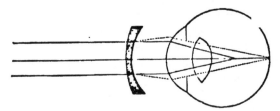

D. Mc. Eye corrected by a Concavo-convex Lens.

NB.—The dotted lines show the correction.

other glasses. Fig. 41 shows (see A, B, C, D) the nearer approach of the visual lines to a perpendicular incidence when they fall on the lateral points of a meniscus and concavo-convex lens than obtains when they are directed to corresponding lateral parts of the surface of a bi-convex or bi-concave lens, and at the same time indicates the wider range of view afforded by the two forms of glasses first named. Fig. 42 (A, B, C, D) shows the differences in the angular obliquity of a lateral ray which falls on the surface of a bi-convex, meniscus, bi-concave, and concavo-convex lens respectively, at an equal distance from the centre of curvature of each.

Duplex Focal, Bi-focal, or Franklin Glasses.—Spectacles with the upper half adapted for looking at distant objects; the lower half for near objects. The line of the lower segment is brought into contact with that of the upper at an angle so that the visual axis, in the different positions required for seeing distant and near objects, meets the surface of each segment at right angles or nearly so.

Thus the upper segment may be made slightly convex to correct H.; the lower half more convex to correct H. combined with Pr.; as the conditions of the segments appear to have been in the glasses first used and described by Benjamin Franklin, after whom they are named; or the upper to correct a full degree of M., as regards distant vision, the lower to correct the M. sufficiently for some relatively near distance, or again to correct a moderate degree of M. as regards distant objects, and Pr. in respect to near

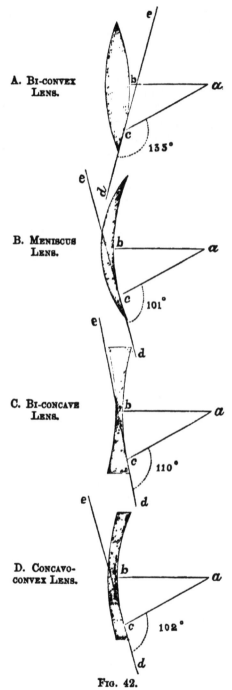

A. BI-CONVEX LENS.   135°

B. MENISCUS LENS.   101°

C. BI-CONCAVE LENS.   110°

D. CONCAVO-CONVEX LENS.   102°

Fig. 42.

ab, central ray; *ae*, ray meeting surface of lens in each instance at distance *cb* from centre; *de*, tangent to spherical surface, at point *c*. The angles *ecd* show the relative obliquity of the lateral rays in the four lenses.

objects. Fig. 43 shows the Franklin principle applied to a meniscus spectacle glass, the lower segment being more convex than the upper and joined to it at an angle suitable for viewing objects near to the eye. When half-lenses are thus employed, their optical centres should be clear of the line at which the upper and lower segments are joined together. They should, therefore, not be constructed by cutting a single glass into two halves to form the corresponding segments in the pair of spectacles, as is sometimes done, but each segment should be cut from a separate lens, and have its own axis complete. When Franklin glasses are thus arranged the wearer is not troubled by the fine boundary line between the two half-lenses ; he acquires the habit of looking above and below it without noticing its presence. In some French lenses the same purpose is occasionally attained by giving the upper half of the spectacle glass a curvature different from that of the lower half, thus combining in one glass the two powers required for distant and near vision without an obvious boundary line between them. They are sold under the name of *verres à foyer double*.

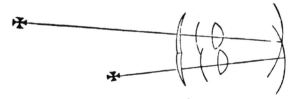

Fig. 43.—Bi-focal Meniscus Glass. The two segments inclined angularly to each other.

Tinted Spectacles.—Plain glasses, or convex and concave lenses, may be tinted in various shades of blue colour, or may be simply darkened (neutral or smoke-tinted glasses), with a view to lessen the effects of glare, as of tropical light, or sunlight reflected from snow or water, or to prevent the irritation caused by light in photophobia from any cause. Orange and red rays appear to be the most irritating rays of the spectrum to weak eyes, and whenever these tints are in excess in the light to which such over-sensitive eyes are exposed, blue-tinted glasses are the most suitable, as they neutralise the orange colour and render the light similar to ordinary diffused daylight. When there is no need to change the colours of objects, and all that is required is to lessen the intensity of brightness of the light reflected from them, glasses of neutral tint are the most fitted for the purpose.

Opticians supply eight degrees of saturation of blue-tinted glasses. A set of specimens obtained from a leading manufacturing optician in London were designated by a series of numbers, the saturation being least in No. 1 and gradually increasing to No. 8. They also supply eight shades of neutral or smoke-tinted glasses, respectively distinguished by the letters A to H, the saturation being least in A and gradually increasing to H.

Eye Protectors.—Spectacles so called because they are chiefly employed to protect the eyes against dust, grit, or in certain trades against particles of metal and other substances which would injure the eyes if they reached their surfaces. They consist ordinarily of spoon-shaped but plain glass, having such forward convexity as readily to cast aside particles on striking them, and equally to protect the eyes from the force of a strong wind when meeting it in front. When they are blue-tinted, and correct in form, they are very useful for protecting weak eyes against the irritating influence of orange-coloured light, and also against the glare of the sun, as in India and other tropical countries. It is important that they should have a suitable curvature for preventing the direct access to the eyes of bright light reflected from below or entering by the sides of the spectacles, for otherwise the eyes will be dazzled and fatigued by the mixture of the untinted rays admitted at the borders of the glasses and the tinted rays transmitted in front. An advantage of the spoon-shaped eye protectors, when compared with some others, is that, while mitigating the effects of strong glare, they allow air to pass freely over the eyes upwards and downwards, and thus are far better than glasses so shaped as to fit close to the orbit, for these heat, relax, and sometimes inflame the eyes by impeding the normal evaporation of their secretions, and by preventing the access of any air to them. They are also better than double spectacles with glass sides, which are objectionable not only on account of heating the eyes but also on account of their weight.

The cheap eye protectors sold in some shops require careful examination, as they are not unfrequently defective in form, so that they do not act simply as a cover and shade to the eyes, but may operate as convex or concave lenses, and may further fatigue vision by causing a certain amount of distortion of objects owing to faulty curvature. They can be easily tested in the same way as has been already described, with regard to convex and concave lenses. Inferior glasses are also not unfrequently coloured unevenly in the grain of the glass, and sometimes contain specks and other imperfections, which are less noticeable than they would be in unstained glass, owing to the fact of the glass being tinted. Such defects are always more or less detrimental and disturbing to vision.

Goggles.—These are also contrivances for protecting the eyes against dust, glare, and other sources of irritation. They are formed of various materials, wire gauze, or a combination of wire gauze and glass, and may be set in spectacle frames or only fitted to the orbits, and held in place by ribbons that may be tied behind the head. The goggles supplied to the troops engaged in the recent military operations in Egypt consisted of two oval flat pieces of blue-tinted glass, set in front of two boat-shaped fine wire-gauze sides. The glasses and the gauze wire were kept in shape by narrow steel frames or borders, and the whole was so fashioned that the edges fitted closely to the bony margins of the two orbits. The eyes were thus completely inclosed within the goggles. Each pair of goggles were connected on their inner aspects by a piece of cord admitting of a certain play, so that the goggles might adapt themselves to eyes at different distances apart; and at their outer edges were connected by a piece of elastic cord sufficiently long to permit it being passed over the ears and behind the head, and to hold them in position when they were worn. When not required for use, one goggle could be placed within the other goggle and the two put together in an oval japanned tin box, $2\frac{1}{2}$ inches long by $1\frac{3}{4}$ inch across, and 1 inch deep, for security. Portability, lightness, and comparative cheapness were thus secured, while they answered the purpose of warding off the grosser particles of sand and diminishing glare.

Prismatic Glasses.—Spectacles fitted with prisms, or with prisms in combination with convex and concave lenses, are occasionally used for various optical purposes; such as to correct slight declinations of the visual lines, whether upwards, outwards, or inwards, and thus to prevent visual confusion from double images; or to relieve asthenopia depending on undue strain of the M. recti interni, as when persons with high degrees of myopia read or work at near distances at which the exertion of accommodation is not parallel with that of the convergence. Their action will vary according to the positions given to the prisms (see Prisms).

Stenopœic Hole.—A very small circular opening in the centre of an opaque metallic diaphragm. The effect of the stenopœic

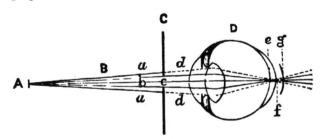

FIG. 44.—STENOPŒIC HOLE.

A, illuminated point of an object; B, cone of light proceeding from A; C, metallic diaphragm with central perforation; D, eye globe; *aa*, marginal rays of the cone, B; *b*, central rays of the same; *c*, stenopœic aperture transmitting *b*; *dd*, dotted lines showing the course that *aa* would take if not obstructed by the metallic diaphragm C, which only transmits the central rays *b*, and so far reduces *e*, a diffusion area formed in front of *f*, the true emmetropic focal point, and at *g* behind the same.

hole, when the eye looks through it, is to exclude the marginal rays of the beam of light, proceeding from each illuminated point of an object, which would otherwise pass through the pupil, and thus to lessen the area of the circle of diffusion on the retina in the case of any eye that is not emmetropic. At the same time that the marginal rays are excluded, a certain portion of the light emanating from an object is necessarily also excluded, so that the luminous impression on the retina is proportionally diminished. Although the actual brightness of the surface of the object looked at may remain the same, the effective brightness is lessened in nearly the same ratio as a transverse section of the cone of rays entering the pupil through the stenopœic hole is less than the cross section of the cone of rays which would enter the pupil if its whole area were free for their admission. Objects looked at through a stenopœic hole, therefore, appear darker, and both the range of vision and the advantages of ocular movements are also curtailed by it. The stenopœic hole is often of use in determining whether deficiency in acuteness of vision is due to ametropia or to some other ocular defect. It is very important for optical purposes that the margin of a stenopœic opening should be perfectly clean and even.

Stenopœic Slit.—A narrow slit in an opaque metal diaphragm. The effect of the stenopœic slit, when placed before the eye, is to limit the entrance and passage of rays of light to one meridian of the eye. It may be conveniently employed in the diagnosis of astigmatism.

Stenopœic Holder.—An appliance provided with a stenopœic slit, so arranged as to be capable of being lessened or increased in width, and adapted for holding concave or convex spherical lenses. The appliance is provided with a suitable handle. It is chiefly employed in the diagnostic examination and in the correction of astigmatism.

Stenopœic Glasses.—Opaque glasses or metal discs with very narrow openings, either in the form of a circle or of a slit, for limiting the transmission of rays of light to the eye. They are used to improve vision, when only a particular portion of the dioptric media is clear, by preventing the disturbance due to light being diffused through partial obscurations of the cornea or of other of the dioptric media; in irregular astigmatism, to restrict the passage of light to a particular portion of the cornea; and also when it is desired to lessen the amount of light entering the eyes, as in tropical countries in mydriasis or cases in which the iris is permanently dilated from any cause, or in which the iris or a large portion of it has been removed by iridectomy.

Strabometer or Strabismometer—An instrument for measuring the linear extent of deviation of the centre of the cornea of an eye, which is displaced laterally from the position it would occupy in normal binocular vision on the two eyes looking directly forward at a distant object, as happens in convergent and divergent strabismus. It consists of a small ivory plate, fitted with a handle, and made suitable in shape for being applied to the lower lid of an eye. At the curved margin of the plate a scale of marks and figures is engraved, indicating distances in millimetres, or parts of inches, on each side of a central mark which is intended to be placed in line with what would be the normal situation of the centre of the cornea. As soon as the instrument is adjusted, the distance of the displaced corneal centre from the line indicating this normal position can be at once read off from the instrument.

Judging Distance.—An important part of the instruction of recruits at rifle drill practice consists in teaching them to estimate correctly the distances of objects by ocular observation alone, that is, without the aid of range-finders, or other means of measurement. If a bullet on being fired from a rifle travelled in a straight path nearly to the end of its course, there would be no need for soldiers to become skilled in judging the distances of objects, or for adjusting their rifles to those distances, for the bullet would strike any object it might meet in its path within the limits of its range ; but since its path, or trajectory, instead of being straight, is curvilinear, it is evident that the bullet in its fall can only strike an object of the height of a man within a certain limited space. If the object be outside this space, either in front of, or beyond it, it will not be hit. Hence the necessity for a correct knowledge of the particular distance of any object that may be aimed at, in order to adjust the rifle to that distance, and so make sure that the object shall be hit by the bullet in its fall.

The power of *judging distance*, as it is called, with approximate accuracy, is therefore essential, for without this knowledge a soldier cannot use his rifle efficiently.

The faculty of judging distance depends upon the capability of properly appreciating the differences in the *visual angles*, or in other words, the differences in the sizes of the retinal images, formed by an object of known size, a man or horse for example, at different distances. The visual angle, and therefore, the retinal image of an object, lessens as its distance increases. In proportion then as the distance of an object of given size from the eye of an observer is increased so will be the diminution of the apparent size of the object ; and conversely, in proportion as the apparent size diminishes, so will be the increase in distance of the object from the eye of the observer.

The acquirement of the art of *judging distances* is effected by causing the men under instruction to note the apparent size and aspect of soldiers placed at certain distances, to observe and familiarise themselves with the appearance of different parts of the figure, limbs, accoutrements, and dress of the men, and to make comparisons between them and the appearances of the same soldiers and objects at various other distances. The appearances of surrounding objects, such as trees, buildings, &c., at different distances, are also impressed on the soldier's mind. The practice is facilitated by observing that certain parts of the men or objects on their persons disappear from view at particular distances; this fact depends, other things being alike. on the visual angles formed by these parts at such distances being too small to permit recognition. In like manner, and from the same cause, the whole object according to its size will disappear from view at some particular distance, even though no other impediments to a clear view of it exist (see ch. vii., pages 155 to 159, for further remarks on this subject).

The difficulty of judging accurately the distances of all objects is greatly increased when they are placed far away from the spectator, yet it is under these circumstances that in using the rifle the need for forming a precise judgment of the distance of any object that may be aimed at is greatest. Hence, the experience derived from repeated practice, and accurate observation and comparison of surrounding objects, as well as of objects themselves, together with very perfect qualities of sight, are essential for a man to become a good marksman at objects at long ranges, when the estimate of the distances at which the objects are placed depends on the marksman's own judgment, as it must often do in military operations on active service in the field. The drop of a bullet increases in curvature in proportion to the elevation for distance given to the rifle, and of course the space within which an object of the height of a man would be liable to be struck is lessened in the same proportion. In firing at distances up to 300 yards at a target 6 feet in height, the bullet does not rise anywhere more than 6 feet above the ground, and hence a man of that height would be hit at any point of the bullet's path within this range; but if elevated to suit a distance of 400 yards, the space within which he could be hit by the bullet in the latter part of its course would be limited to 110 yards; if at 800 yards, the limit of the dangerous space would be lessened to 45 yards; if at 1,000, there would only be a space of about 20 yards within which an object 6 feet in height would be liable to be struck. It is obvious that at such long distances, and under such reduced visual angles as objects of the height of an infantry soldier or trooper would subtend at them, the eye must be capable of getting retinal images quite clear of circles of diffusion in order that the objects may be distinguished plainly.

A certain amount of assistance in judging the relative distances of objects, when several are regarded in succession, is derived from the sense of the varying efforts of accommodation, and also from a consciousness of the consentaneous movements of the two eyes in changing their positions for the purpose of so adjusting their visual lines that they may meet together in the same points of the different objects under observation.

It is not to be forgotten that, irrespective of the particular quality and condition of the dioptric media of the eyes of any individual observer, the character of the retinal image of an object and the acuteness of the visual impression produced by it, will be modified very materially by the degree of illumination of the object, the position of the sun in regard to it, the character of the background, particularly its amount of contrast with the object, and the state of the atmosphere between the object and the observer in respect to its transparency and refractive stability.

Aiming.—Before a recruit is permitted to practise firing his rifle either with cartridge or ball, he has to go through a course of instruction in 'aiming drill,' or the mode of aligning the 'sights' of the rifle on the object aimed at; in other words, in so holding his rifle that the parts of it named, and the object to be aimed at, all simultaneously occupy positions in the line of vision. There are two sights attached to the rifle, technically called the *backsight* and the *foresight*; a slide on the former being capable of being raised to various heights, while the foresight is fixed. The recruit is taught to look steadfastly with his right eye along the bottom of a notch in the backsight in the direction of the projecting top of the foresight, which, in turn, is to cover the centre of the mark towards which the eye is directed. The backsight has to be kept perfectly upright, and the slide upon it to be so placed as to adjust the rifle in respect to elevation to the special distance of the object ordered to be aimed at. The recruit is instructed to fix his eye steadfastly on this object, not on the barrel of the rifle or the foresight, keeping his left eye closed, while he brings the top of the foresight in line with the object through the notch of the backsight. If more than the tip of the foresight is brought up into the alignment, a little additional elevation of the rifle is the result. This instruction is carried on for increasing distances until the soldier has become perfect in aiming, for the difficulty of aligning the foresight accurately increases as the distance of the object increases.

Medical officers are occasionally appealed to by musketry instructors on account of their meeting with difficulties in getting some men to adjust the sights of the rifles so as to bring them into true line with the object, and from doubts arising in their minds as to whether the men concerned are not suffering from defective eyesight. It is well, therefore, for medical officers to be prepared with a knowledge of the visual conditions which may facilitate or hinder men from taking a correct aim.

When a Martini-Henry rifle is brought to the front of the right shoulder of a soldier in the standing position the distance of the notch in the slide of the backsight is about 15 inches, and of the 'tip' or 'fine sight' of the foresight about 39 inches, from the right eye. When an object, such as the bull's-eye of a target at a distance of several hundred yards, is aligned with these sights of the rifle, it is obvious that an eye, if it tried to do so, could not see the three different objects with equal distinctness at one and the same time. The recruit, therefore, is rightly instructed to fix his eye steadily on the mark he is to aim at—not on the sights nor the barrel of his rifle, which, although not directly looked at, can yet easily be brought into the line of vision with the bull's-eye of the target. As soon as the sights and object are brought together into true line, the recruit is rendered conscious that they are so, by the simple fact that the foresight, projected through the notch of the backsight, intercepts some of the rays of light reflected from the bull's-eye of the target which would otherwise reach his eye ; and he learns that his rifle, if its elevation be duly adapted to the distance, is then in the right direction for a missile discharged from it to strike the target. He does not look at, nor recognise with distinctness, either of the sights of the rifle ; he only sees clearly the bull's-eye of the target, and at the same time observes its partial obscuration owing to the intervention of the foresight.

Monocular Aim.—The recruit is taught to close the left eye when aiming at an object with his rifle. By so doing he confines his view to the object and a comparatively limited field around it, while he finds it easier to align the sights of the rifle upon the object. It is doubtful whether this plan is advantageous under all circumstances, especially when the object aimed at is not a fixed one at a given distance, as it is at target practice.

Binocular vision gives a more vivid impression of an object from both eyes being simultaneously centred upon it, while it enables the observer to obtain a better notion of its distance and form, as well as of the direction and rate of its motion, when the object is a moving one, such as an enemy running or riding in an open landscape. There is also a further advantage gained under some circumstances from binocular vision, through an interchange of sympathetic action between the two eyes. This is independent of the ' searching motion of the eye,' so fully described by Sir C. Bell, the purpose of which is to obviate the effects of retinal exhaustion when vision is intently concentrated on a particular spot in the field of view. If the gaze of a single eye be fixed on a distinct point of an object, the other eye being closed, notwithstanding the ' searching motion ' above referred to, the point quickly fades from view owing to retinal exhaustion, whereas, if both eyes be open, a clear view of the point gazed upon can be maintained for a much longer time. As the sensibility in the region of the macula lutea of one eye is becoming exhausted, the necessary energy is provided by the other eye, and this alternate ebb and flow of retinal power between the two eyes enables the duration of the maintenance of the intent gaze on the object to be considerably prolonged.

But a man cannot align the sights of the rifle upon a given distant point, as when at target practice, when both eyes are open and visually directed towards that point. He can only do so if he has the power, although both eyes are open, of arresting the use, as it were, of the one eye while the other is taking aim ; that is, of mentally suppressing note of the image formed on its retina. When one eye only is used, the sights of the rifle, or such parts of them as are not covered by other parts nearer to the eye and the object aimed at, are all in the visual line, and consequently together form an image on the most sensitive part of the retina, the fovea centralis. When both eyes are used, and the object aimed at is a long distance off, the visual lines are nearly parallel, and the distant object is imaged on the fovea centralis of each eye with binocular advantages; but, under these circumstances, the images of such near objects as the sights of the rifle are formed on parts of the retina outside the region of the fovea centralis, and are consequently less distinctly visible than they are when they are in the course of the visual line under monocular vision. For either of the sights of the rifle to be seen distinctly the visual lines must have a very different relative direction ; the two eyes must converge on the sight concerned, and the visual lines be inclined toward it at an angle corresponding with the proximity of the object.

These facts may be illustrated by two simple experiments with a rifle.

When a distant object is regarded with a single eye, the other being closed, the rifle and sights aligned upon the object are seen with a certain amount of distinctness ; if, without any change of position, the other eye be now opened, and both eyes are directed on the distant object, the distinctness of view of the rifle and its sights, especially of the backsight and parts approaching the stock, is at once lessened. The rifle is no longer in the line of sight; it is outside of it. Or if an object be aimed at from a rest, and the rifle be placed in a straight line midway between the two eyes while they are fixed on the object, the object may be seen distinctly, but the barrel, and, of course, its sights with it, appear very shadowy, because the rifle does not coincide with either of the visual lines of the two eyes; it is between them. The backsight, if elevated, will probably appear double under such conditions.

If a rifle be not at disposal the same facts may be illustrated simply by the hand alone. If the knuckle of the right hand be applied against the lower border of the right orbit, and the forefinger extended so as to cover a portion of a distant object, while the left eye is closed, the finger and object will be seen together with more or less distinctness, being together in the line of the visual axis; if now the left eye be suddenly opened, the finger will disappear from view, while a more vivid perception of the distant object, and a more extended view of its surroundings, will be obtained. After a short time, if the mind be intently directed to

the purpose, the use of the left eye may be suppressed although the
eyelids remain open, when the finger and the object to which it is
directed will be again seen together with a certain amount of
distinctness. This will be more easily accomplished by persons who
are in the habit of ignoring the image formed in one eye, as
happens with surgeons who are accustomed to use the ophthalmoscope, microscope, and other such optical instruments, or as many
sportsmen do who are in the habit of firing with both eyes open.
Consideration of the facts just mentioned will explain why musketry
instructors find that men who cannot shoot with an approach to
accuracy at a mark when both eyes are open can shoot fairly with
one eye closed, although there is no difference in the two eyes as
regards refraction or visual power.

As the central line of the notch of the backsight and the slender
ridge or tip of the foresight are very fine objects, and very accurate
recognition of them is necessary in ensuring their perfect alignment
with a distant object, it seems probable that in taking a precise
aim they are each looked at in rapid succession, in other words,
that an alteration of accommodation takes place from one to
the other object, but so rapidly that the view of them is practically
simultaneous, so that the alternate adaptation of vision to the different distances escapes notice. If such a rapid visual adaptation
be essential for ensuring a perfect aim, as seems likely, it is obvious
that a person who is either hypermetropic or presbyopic in any
considerable degree would experience special difficulties in taking
a correct aim with a rifle at a remote object; for he could not see
the foresight of the rifle with sufficient definition and alertness.
A hypermetropic eye, even at the recruiting age, if the hypermetropia amount to -2 D, sees a small object, such as the tip of the
foresight, so very hazily, that accurate alignment, unless the optical
defect is corrected by a suitable lens, is out of the question. Also
for a myopic eye, if the myopia be of corresponding amount, or
equals $+2$ D, alignment is rendered very difficult because, although
the tip of the foresight is less obscured than it is with the hypermetropic eye, there is still such reduplication and haziness of the
distant object as to make the amount of the foresight brought up
into the alignment and the relative positions of the sights and the
object all very uncertain. If the ametropic condition be complicated with astigmatism, a certain amount of alteration in form
and apparent change of dimensions are added to the haziness of
view already mentioned, and the difficulties in the way of accurate
alignment are increased.

Strengthening the Eye for Shooting at Long Ranges.—It is
laid down in the book of Musketry Instruction that ' it cannot be
too strongly impressed on every man that to shoot well at long
ranges he must train and strengthen his eye by looking at small
objects at long distances ' ('Musketry Instruction,' 1884, par. 77,

p. 84). A study of the conditions on which accurate vision of distant objects depends sufficiently shows that if an eye be emmetropic, neither the eye nor the quality of vision for such remote objects admits of increased strength or improvement; if an eye be myopic from natural formation, there is no reason for believing that exercise on distant objects will lessen the degree of myopia or its effects, or, if it be hypermetropic, make up for the deficiency otherwise than by the ordinary exertion of accommodation. But other conditions on which accurate shooting also depends may be developed, and these belong more to the brain than to the eye. Minute features and peculiarities of a given object and its surroundings, which are unnoticed at first view, are rendered familiar and attract critical attention under close and repeated observation, and relations of size and distance of objects become more correctly estimated. It is by these means that painters perceive particular features of a face or landscape, and sailors see and distinguish objects on a distant horizon, although the images of those objects are not pictured with any more precision in their eyes than they are in the eyes of others with the same quality of sight who, however, fail to recognise their presence in the field of view. The artist and sailor learn to observe with more precision, and they acquire by practice the art of distinguishing and rightly interpreting differences, often very slight, in form and colour, or contrasts in light and shade, which are unnoticed by a less observant or less experienced spectator, notwithstanding he may be equal in strength of eye and acuteness of vision. It is not the strength of the eye which is developed so much as the faculty of observation, and it is to this the attention should be chiefly directed.

The recent introduction of figure targets as objects to be aimed at in the range practice at musketry instruction will probably assist men in acquiring an increased facility of judging distance. The different distances at which the 6-feet figure is fired at along the range must tend to educate the eye and to enable it to estimate rightly any particular distance of the figure through the familiarity which will gradually be obtained with the different appearances presented by it at the various distances it is fired at.

Abbreviations.—The following abbreviations are occasionally employed in this manual: Acc., for Accommodation; Ambl. for Amblyopia; Em. for Emmetropia; Emc., Emmetropic; M., Myopia; Mc., Myopic; Hm., manifest Hypermetropia; Hl., latent Hypermetropia; H., absolute Hypermetropia; Hc., Hypermetropic; Ast., Astigmatism; C., Convergence; Cyl., Cylindrical; V., Vision, or acuteness of vision; Pr., Presbyopia; Prc., Presbyopic; P., proximate or nearest point of distinct vision; R., remote, or most distant, point of distinct vision; Lt., Left; Rt., Right; Ref., Refraction; Sn., Snellen's types; Sph., Spherical; D., Dioptric; 1 m., one metre; 1 cm., one centimetre; 1 mm., one millimetre.

Symbols.—The following symbols are also found useful for the purposes of abbreviation :— 1′, one foot; 1″, one inch; 1‴, one line; ∞, infinite distance, or infinity; ◯, combined with; +, convex, and ◡, concave, when applied to lenses.

CHAPTER II.

Vision according to varieties of Focal Adjustment—-Chief Varieties—EMMETROPIA —-Definition— Optical Conditions—Farthest and Nearest Points of Distinct Vision—Diagnosis—MYOPIA—Definition—Farthest and Nearest Points of Distinct Vision — Optical Conditions — Causes — Symptoms — Counterfeit Myopia—Association of M. with Strabismus—Diagnosis—Expression of Degree of M.—Determination of Degree—Correction—Exaggeration of M. from Spasm—Over-correction - Extension of Reading Distance in High Degrees of M.—Test in Military Practice for High Degree of M.—Influence of M. on Military Service — HYPERMETROPIA — Definition — Optical Conditions — Farthest and Nearest Points of Distinct Vision—Causes—Symptoms—Association of H. with Strabismus—Manifest and Latent H.—Subdivisions of H.— Diagnosis—Expression of Degree of H.—Determination of Degree—Example —Explanation—Correction of H.—Influence of H. on Military Service— ASTIGMATISM—Definition—Optical Conditions—Causes—Symptoms—Retinal Images in Ast.—Amount of Ast.—Measure of Amount—Kinds of Ast—Diagnosis—Varieties of Regular Ast.—Examples—Determination of Principal Meridians in Ast.—Their Refractive Conditions—Correction of Regular Ast.— Examples—Irregular Ast.—Influence of Ast. on Military Service.

Chief Varieties of Focal Adjustment. GENERAL REMARKS.— Measure of focal distance being the chief feature of those conditions of vision which depend upon the optical adjustment of the eye in a passive state of rest or the state in which there is no active exertion of accommodation, all varieties in this respect may be classed under three heads or natural divisions. They are expressed by the following terms: 1. EMMETROPIA, focus in correct measure, the refractive power of the eye being normal; 2. HYPO-METROPIA, under measure, or Brachy-metropia, focus short in measure, the refractive power of the eye being in excess; 3. HYPER-METROPIA, focus beyond measure, the refractive power of the eye being deficient in degree. But, as regards No. 2, Hypo-metropia and Brachy-metropia, the old classical term MYOPIA, derived from the habit of short-sighted persons partially closing or nipping together their eyelids, is so familiarly known, and is so convenient for avoiding mistakes from similarity in sound between the Greek derivatives signifying under and over, that its use has been universally continued.

The diagrams which follow, figs. 45, 46, 47, serve to illustrate in outline the three different refractive conditions, or varieties of focal measurement, just named.

AMETROPIA, not in measure, is a term which simply signifies that the condition of emmetropia does not exist; it therefore comprehends the condition of myopia as well as that of hyper-metropia.

ASTIGMATISM is a variety of disordered focal adjustment of a composite nature, in which different degrees of refractive power

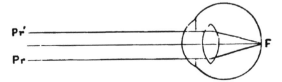

FIG. 45.—EMMETROPIC EYE. Parallel rays focussed on retina.

exist in the same eye in different meridians. The principal elements of Astigmatism, however, when regarded separately, belong to one or other of the three leading optical divisions above named.

The terms ISOMETROPIA, with its adjective form *isometropic*, are sometimes used to express briefly that both eyes of an individual

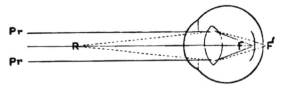

FIG. 46.—MYOPIC EYE. Parallel rays focussed in front of retina.
R, remote point of distinct V., from which divergent rays are focussed on the retina.

are equal in refractive quality and power; ANISOMETROPIA, and *anisometropic*, that, although similar in refractive quality, they differ in their degrees of that quality ; ANTIMETROPIA, that they are dissimilar in quality, as when one eye is emmetropic but the other ametropic ; or that their refractive qualities are both defective and

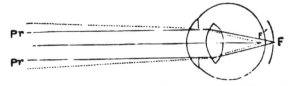

FIG. 47.—HYPER-METROPIC EYE. Parallel rays focussed behind the retina.
Convergent rays focussed on retina.

at the same time opposite in kind, as when one eye is myopic and the other hyper-metropic.

The varieties of focal adjustment which have been named refer to the refractive quality of the eye when it is in a state of complete repose ; but the focal adjustment of every eye under normal conditions may be more or less changed by the exercise of

a special function named ACCOMMODATION. The various refractive
conditions, as well as the function of accommodation, will now be
described more fully.

EMMETROPIA.

Definition.—Normal-sightedness, in respect of the refractive
power of the eye in a state of repose—that is, when it is not exerting any of its accommodatory power.

Optical Conditions.—The dioptric media of the eye are so
adjusted that, by means of their combined refractive qualities, rays
of light emanating from distant objects—that is, objects at a distance of 6 m. (20 feet) and upwards—and falling upon the eye practically as parallel rays, are brought accurately to a focus upon the sentient layer of the retina. The eye is in true measure. In other
words, the retina is so placed as regards measure of distance that
it is precisely in the plane of the principal focus of the dioptric
media of the eye; the two distances, that of the retina and that of
the principal dioptric focus, are in perfect harmony (see fig. 45).

Farthest Point of Distinct Vision.—Infinite distance, or the
distance of the farthest visible objects—the fixed stars for example.
However great the distance of an object may be, so long as the
other conditions necessary for vision, viz. sufficient size and illumination, are preserved, a perfectly defined retinal image of it will be
formed by the emmetropic eye.

Nearest Point of Distinct Vision.—Varies with age, and power
of accommodation (see ch. iv.).

Diagnosis of Emmetropia may be established subjectively: (1)
by types, (2) by spectacles, and (3) objectively by the ophthalmoscope.

1. *By Types.*—The emmetropic eye, other conditions being
favourable, can read Snellen's types at the indicated distances with
clearness and facility.

2. *By Spectacles.*—Vision of distant objects is not improved
either by a convex or concave lens as a hyper-metropic or myopic
eye respectively would be, but, on the contrary, is deteriorated in
proportion to the increase in power of the lens applied to the eye.
If, however, a weak concave lens be applied to an emmetropic
eye, its deteriorating effect may be counteracted by an exercise of
Acc. on the part of the person under examination, especially if he be
young. Printed letters of small type can be read at a distance of
10 inches from a + 4 D lens, placed in front of the eye at the distance of the anterior focus of the eye, about half an inch from
the cornea, for the rays of light coming from the print are thus
caused to fall on the eye as parallel rays. The farthest point of
distinct vision of the emmetropic eye with a + 4 D lens before it
is at 10 inches.

Occasionally, as age becomes advanced, the subject of vision which has appeared to be emmetropic finds that he derives some advantage in looking at distant objects from the use of convex glasses of moderate degrees of power. This is a change which Professor Donders designated 'acquired hyper-metropia,' and in his great work on *Anomalies of Refraction* he has given a diagram showing from various observations the course of this change. According to this diagram the change begins at middle life, and at 80 years of age the acquired H. amounts to a deficiency of refractive power equivalent to 1·5 D, or a lens having a focus of 26⅔ inches. It is questionable whether in such instances the so-called acquired condition of H. is really due to senile changes, or whether the eyes under observation have not always been hyper-metropic although the H. has not been detected, owing to its being supplemented by a certain amount of accommodation. Practically, however, whatever the explanation may be, whenever the condition exists, the fact that vision for distant objects is improved by convex lenses takes the eyes under observation out of the category of emmetropia and puts them in that of hyper-metropia.

3. *By Ophthalmoscopic Observation* (see page 106).

MYOPIA.

Definition.—Syn. : Short-sightedness. Near-sightedness. The measure of distance of the principal focus of the dioptric media of the eye is under the measure of the principal axis of the eye, and the focus, therefore, does not reach the plane of the retina. The refracting power of the eye, when it is free from all exercise of accommodation, is in excess by comparison with that of an emmetropic eye; so that parallel rays from remote objects are brought to a focus in front of the retina. Only divergent rays are focussed upon the retina; vision is, therefore, acute as regards objects at and within a certain limited distance, while it is indistinct as regards more distant objects (see fig. 46).

Farthest Point of Distinct Vision.—A fixed limited distance in front of the eye, the measure of which varies in different persons according to variations in the degree of myopia.

Nearest Point of Distinct Vision.—This also varies in distance with the degree of myopia, together with the amount of Acc. that can be exercised, but it is always nearer to the eye than it is in a person of emmetropic vision at a corresponding age, or with a corresponding amount of accommodatory power.

Optical Conditions.—To produce a myopic condition of vision, either the antero-posterior axis of the globe of the eye must be prolonged beyond the distance of the principal focus of the refracting media (and this is the general cause in simple myopia), or the refracting qualities of the dioptric media must be increased out of proportion to the length of the eyeball, irrespective of exercise of accommodation. In practice it is most convenient to regard the myopic state as depending on an excess of refractive power in the media of the eye, and to take no note of the increased length of the antero-posterior axis of the eyeball.

Causes.—Hereditary conformation of the eye is the most common source of myopia, it being frequently transmitted by parents, and seen to prevail in many members of the same family. Originating in the majority of cases in artificial and avoidable causes, it becomes transmitted to children, who, in turn, from similar causes, are in many instances rendered myopic in a more marked degree than their parents.

Morbid changes, of which distension, attenuation, and protrusion posteriorly of the sclerotic coat, together with an atrophic condition of the choroid coat, are the most conspicuous features, constitute its secondary causes. This morbid state may either be general, the posterior hemisphere of the eye being elongated in all directions; or, as frequently happens, it is local, the eye being more elongated in the direction of the antero-posterior axis than in other directions (staphyloma posticum).

Conditions in which the convexity of the cornea is unduly increased are also causes of M.

Another cause is long-continued convergence of the eyes and ocular over-exertion at near objects, particularly by children. Under such conditions a certain amount of compression of the anterior portion of the eyeball is produced, and a strain is thrown on the tunics in the neighbourhood of the posterior pole, giving a tendency to elongation of the globe, and going on eventually, if the origin of the mischief be not stopped, to staphyloma posticum. Neglect of adopting proper measures for preventing or remedying the excessive strain just named is a very common source of aggravation of existing myopia, especially when it is associated with a weak state of bodily health. The M. becomes progressive, and under additional conditions of intense light, heat, and glare, such as men are exposed to in India and other tropical climates, the M. is apt to become complicated with serious disorders of the retinal and choroid tunics.

Under similar conditions of continued strain the lens may lose some of its subjectiveness to change of form, or, from being so constantly accommodated for a very near point, the accommodatory apparatus in an eye that is Emc., or nearly so, may cease to be able to relax itself sufficiently to allow parallel rays to be brought to a focus upon the retina. In this last case the parts concerned in Acc. would be primarily in fault, but acquired M. would become one of the results.

Certain neurotics, as the extract of the Calabar bean, by exciting contraction of the ciliary muscle and constrictor pupillæ, temporarily change an Emc. into the condition of a Mc. eye.

Symptoms.—In uncomplicated M. near objects up to a limited point are seen quite clearly. Hence the common name of the affection—near-sightedness. The nearness with which objects can be brought to the eyes enables minute letters to be read, because under such conditions their retinal images are of sufficient size, and sufficient light is reflected from them, for recognition. The circumstances named enable myopes to decipher small print in a light so relatively dim that emmetropes, who would have to hold the print farther off, would fail if they endeavoured to accomplish the same task. More remote objects appear to myopes to be misty, enveloped in a haze, and are consequently only indistinctly visible by them. When distant objects are highly illuminated or brightly coloured, they often appear solid toward their central parts, but attended by a shadowy apparition, or spectral reduplication, of their outlines. The reason that objects near to the eye are naturally seen by a myopic person with precision is that the pencils of rays proceeding from all the illuminated points of these objects impinge on the eye more divergently than they do from distant objects, so that, the increase in divergency being rendered proportionate to the over-measure in the convergent quality of the eye relatively to the length of its visual axis, the rays are brought to a focus upon the retina, and proper images are formed there. At the farthest limit of distinct vision, the over-refractive quality of the Mc. eye is just balanced by the divergency of the rays proceeding from the object looked at at that limit; within that distance, as the rays from objects become more divergent, the accommodatory function is called into play to supplement the refractive quality in order that the eye may still get clearly-defined images on the retina. The reason why an object beyond the far point of distinct vision appears ' blurred,' or surrounded by a halo, is that after the rays proceeding from it have been brought to a focus short of the retina, they cross each other, proceed divergently, and become spread out upon the retina. An image of the object is then formed, but as the pencils of rays proceeding from the object. instead of being focussed to corresponding points on the retina, are spread out in successive circles, the image is necessarily confused and indistinct. The rays which proceed to form the image consist of a number of sections of diverging cones, and these sections overlap each other in its formation. The circular outline of the image is determined by the form of the pupil through which the rays originally passed into the eye. The diffusion is greatest toward the circumference of the image, because the rays of the conical pencil become more divergent in proportion as they approach its peripheral limits. These scattered rays, for sake of brevity, are called ' Circles of Diffusion.' The farther an object is removed from the eye the greater is the diffusion on the retina of the rays proceeding from it, for each pencil of rays from the object is brought to a focus at a greater distance from the retina, and the cone of rays proceeding from this focus has, in consequence, a longer axis and a base of proportionally longer diameter where it is intercepted by the retina, than it would have if the ocular focus from which it started were nearer to the retina. The habit which myopic persons have of partially closing their eyelids, and thus narrowing the interpalpebral fissure, when looking intently at a distant object, arises from an instinctive endeavour to prevent some of the peripheral rays of light reaching the eye, and so lessening the circles of diffusion upon the retina.[1]

When M. is due to the natural form of the eye, and morbid changes have not been induced in the organ by unhygienic mismanagement, the M. does not lessen in degree with advancing years in all, though it may do so in some, instances. In illustration of this fact I may mention that a relative of mine at eighty-five years of age used the same concave glasses he had worn as a youth and assured me they rendered distant objects as clear to him as they had ever done. He had never worn glasses when at such near occupation as reading and writing, and as his M. amounted to about 4 D he never required correction for presbyopia, though he became rather dim-sighted a year or so before his death in his eighty-ninth year.

Counterfeit or False Myopia.—It is necessary to be on guard against counterfeit M., or a condition which may pass for M. but is really only a semblance of it due to irritation and spasmodic action of the parts concerned in the exercise of accommodation. Prolonged work at small objects, as in etching, long-continued reading small and defective print, and other similar near work, will sometimes not merely cause an exaggeration of M. when it already exists, making the patient appear to be more myopic than he really is, but in occasional cases will induce an imitation of a considerable degree of M., particularly in eyes that are weak and excitable from any cause, although no M. may be present. A person may exhibit the condition referred to, and may even have a Hc. formation of the eye. The question of the true refractive quality of an eye under such circumstances can only be solved after the power of accommodation has been paralysed by atropia.

It has been proved by experienced observation in countries where conscription is in force that young persons with good ac-

---

[1] I was once informed by some myopic candidates for commissions in the army that they had been advised by their trainers to apply Calabar gelatines to the eyes shortly before appearing for physical examination. The intention was to produce contraction of the pupils, and so to lessen the diameters of the circles of diffusion. On the other hand, the use of eserine to produce a condition simulating that of myopia by lessening the size of the pupil and shortening the limit of distinct vision, is said to have been not unfrequently resorted to by persons on the Continent, for the purpose of avoiding military service; and on this account, in the military regulations of some countries, it is ordered that the ocular examination of conscripts in special cases shall only take place after the accommodation has been paralysed by the employment of atropine. A simulated, or exaggerated, myopic condition artificially produced by the use of eserine would thus be counteracted.

commodation wishing to avoid military service have been able by voluntary practice to acquire the power of reading print, both distant and near, with concave glasses of higher power than was required for the neutralisation of their real degree of M. They thus succeeded in counterfeiting an excess degree of M. sufficient to cause their exemption from service, when the disabling degree was determined by the use of concave glasses of certain degrees of strength.

Association of Myopia with Strabismus. —M. is not unfrequently found to be associated with strabismus. The squint is occasionally, but rarely, convergent, in cases where the internal recti muscles have become disproportionately developed from constant use owing to the approximation of objects for distinct vision, and so do not admit of being relaxed in proportion to the external muscles. It is much more frequently divergent. This prevalence of divergent squint with high degrees of M. is usually thus explained. As the Mc. condition leads the patient to bring small objects near to the face to be seen clearly, strong convergence of the visual axes is also necessary in order to ensure binocular vision; and thus a strain is thrown upon the internal recti muscles, which they are unable to maintain beyond a limited period of time. Moreover, when the degree of M. is so high that the distant point of distinct vision is only a few inches off in front of the eyes, no accommodatory exertion is required as there would be at a similar distance in the case of an Emc. person. But at so near a distance a person cannot avoid exercising instinctively some amount of Acc. in association with the convergence of the visual lines, and the result is that the refractive state of the eye is added to, and the distant point of distinct vision brought still nearer. The M. is, as it were, for the time increased in degree. A greater strain is thrown upon the muscles of convergence, and the desire is naturally created to relieve this strain, and at the same time to get clearer images of objects farther off than the short myopic distant point. Should an increase in the length of the antero-posterior axis of the eye gradually result, increased M. will be associated with it; and the difficulties of the internal recti muscles in effecting the necessary movements of the eyeballs, and maintaining convergence, will be still further augmented. Under these several circumstances the ocular adductors get so over-fatigued that they cease to act in true concert, and confused, or double vision is liable to result. When this happens, one eye, probably the stronger and more acute, will be directed to the object in view, while the other eye at the same time will deviate a little outwards; a less central, and therefore less sensitive portion, of the retina of the deviated eye will then be brought in line with the object, and thus the mental recognition of its image will be avoided with more ease. Or the eye not in use may deviate still further outwards, and receive only the rays coming from distant objects, when, from the myopic state of the organ, very diffuse and easily disregarded images will result, and so again visual confusion will be prevented, while, at the same time, the aching from asthenopia and the uneasiness connected with the stretching of the posterior tunics of the globe under excessive convergence will be averted. Under both these conditions vision will become practically monocular, not binocular. The strabismus may be temporary in its nature, occurring only when near objects are fixedly regarded, or it may be rendered permanent by repetition of the circumstances just explained, and continue also when distant objects are looked at.

Diagnosis of Myopia.—This may be established subjectively : (1) by external signs, (2) by types, (3) by lenses, (4) by + 4 D spectacles, (5) by correction, and (6) objectively by the ophthalmoscope.

1. *External Signs.*—The myopic eye usually presents some peculiar characters indicative of its condition. It is prominent, or even appears to protrude ; the pupil is usually contracted, and the constant nipping together of the lids produces a noticeable appearance of the parts immediately surrounding the eye. The existence of divergent strabismus, whether observable only when the eyes are directed to very close objects, or constant, as already explained, may be a further diagnostic sign of myopia.

2. *By Types.*—The Mc. eye, if no other defect exist and the M. be moderate, not above 1·50 D, will be able to read No. 1 and also No. 2 of Snellen's types at the proper distances, but will not be able to read the larger types at the distances indicated by the figures placed above them. If the degree of M. be higher than 3 D, then No. 1 of Sn. will not be read at quite the distance of one foot. The farthest distance at which the smaller types can be distinctly read will indicate the probable degree of the M.

3. *By Lenses.*—This method, sometimes referred to as Donders' method, is very simple when a complete trial case of lenses is available for use. The subject being in position before Sn. types No. XX., or D = 6, a weak +, and then a weak − lens is placed before the eye to be examined, the other being shaded. Should the + lens be found to make V. worse, while the − lens improves V., this result establishes the diagnosis. If the latter be a − 1 D lens, lenses of gradually increasing power are applied, until normal V. or a near approach to it is attained. The weakest of the series of lenses which gives clear vision of the distant types changes the eye into the same condition as an Emc. eye, and gives the measure of the degree of M.

4. *By the + 4 D Spectacles.*—When the + 4 D spectacles are worn, as before described, one eye being covered, and small type is placed before the uncovered eye at a distance of 10 inches from the lens, it is found that the type cannot be read, but on bringing the type nearer to the lens it becomes legible. The farthest point

of distinct vision of the Mc. eye, with the $+4$ D lens before it, is, therefore, short of 10 inches. In examining the refractive conditions of eyes with spectacles, the patient should stand with his back to the light, which should so fall on the print as to illuminate it thoroughly. The print should be advanced gradually towards the lens, a rigid inch measure being held at the same time horizontally by the side, and the distance of the first point at which the print becomes clearly defined should be carefully noted. In shading the eye not under examination, the lids should not be closed by pressure of the fingers, but the eye should be simply covered by the hand. If the eye be pressed upon, some minutes must elapse before it will recover a suitable condition for optical examination.

5. *By Correction.*—When the true degree of uncomplicated M. is ascertained, the proper concave lens for that degree will *completely* correct the abnormal condition, and the person under examination will be able to read Sn. 6m. at a distance of 20 feet.

It must be remembered, however, that it is not always easy for a myopic subject to decide with precision the lens which exactly neutralises his M., or, in other words, the glass which suffices to give him distinct vision for distant objects without over-correction. This may arise from several causes. The glass may be stronger than necessary, and the myope unconsciously exert Acc. to neutralise the artificially produced deficiency of refraction, just as a hypermetropic person would do, or from the presence of a certain amount of amblyopia, and defective acuteness of visual observation, there may not be sufficient difference in the apparent distinctness of the objects, when two or three concave glasses of approximate powers are used, for the myope to be able to appreciate the difference which really exists. If, however, a correcting glass is found which enables a myope to read No. 6 m. Sn. at 6 m., or 20 feet, while a weaker or stronger glass does not enable him to do so with equal facility, the myopia is probably just corrected, and not over-corrected.

6. *By the Ophthalmoscope* (see p. 106).

Expression of the Degree of Myopia.—The degree of M. may be either expressed by the power of the concave lens whose principal focus is situated at the same distance from its centre as the remote point of distinct vision of the Mc. eye under observation, or by the power of the convex lens which represents the amount of the excess of refracting power of the eye relatively to emmetropia. The latter mode of expression is the more convenient in practice. M. $= 2$ D then signifies that the excess of refracting power in the eye equals the power of a $+2$ D lens, and that this excess will be neutralised or corrected by a -2 D lens. The distant point of distinct vision of the eye with myopia $= 2$ D will be at the same distance as that of the principal focus of the -2 D lens, viz. at 20 inches' distance.

Determination of the Degree of Myopia.—When the existence of M. is established, its degree may be determined by simply finding the distance of the remote point of distinct vision of the eye under observation, or, in other words, the farthest point at which small objects can be seen with perfect definition, and beyond which they begin to appear blurred. The distance of the remote point of distinct vision agrees with that of the principal focus of the lens which represents the degree of M. But for reasons explained below (see footnote) it is more convenient in military practice to determine it by the $+4$ D spectacles. The distant point of distinct vision is accurately ascertained after the $+4$ D lens has been placed before the eye at the usual distance of spectacles, viz. half an inch in front of the cornea. Inverting the distance so found, and deducting from it so inverted the power of the lens added to the eye, viz., $\frac{1}{10}''$ the difference gives the degree of myopia.

Example 1.—Suppose the distant point of distinct vision of the myopic eye with the $+4$ D or $+10''$ lens before it, is found to be $6''$, then $M. = \frac{1}{6} - \frac{1}{10} = \frac{1}{15}$. In calculating by the metric system the distance of the remote point must be converted into dioptrics, and the dioptric power of the lens added to the eye, viz. $+4$ D, must be subtracted from it. The difference gives the degree of M. Thus in the example just taken, as 6 inches is the focal distance of a $6\cdot66$ D lens, then $6\cdot66$ D-4 D$=2\cdot66$ D, the degree of M.

Explanation.—The following will explain the process more fully. Let $x=$ the refracting power of the eye under examination; let $u=$ the refracting power of an emmetropic eye; $+4$ D$=$ the power of the lens to which the eye has been subjected; and $+6\cdot66$ D $=$ the power of the lens which would give the ascertained distant point of $6''$ if the eye were emmetropic. The excess of refracting power in the example given is therefore obviously equivalent to the difference between a $6\cdot66$ D and a 4 D lens. This excess can only be in the eye itself, and its refracting power must evidently be reduced to a corresponding degree to bring it to a par with an Emc. eye.

Therefore $x-(6\cdot66$ D-4 D$)=a$; or $x-2\cdot66$ D$=a$; or $x=u+2\cdot66$ D. M., the excess of refracting power over that of an emmetropic eye, is$=2\cdot66$ D, or is equivalent to a lens of 15 in. focus.[1]

---

[1] *Reasons for Adopting the Method described in the Text.*—The use of the con stant $+4$ D or $+10''$ lens for examination of ocular states of refraction is adopted in preference to other methods for military purposes, because it is equally applicable to the determination of Emmetropia, Myopia, and Hypermetropia, and of the degrees in which the two latter conditions exist; because medical officers cannot usually avail themselves of regular series of lenses for conducting such investigations; because the mode of observation can be easily and quickly learned; the observations can be conducted within a moderate range of distance, such as can be obtained in any ordinary room; and further, since the trials are usually

The Correcting Lens.—The exact excess of converging power having been determined, it is corrected by a lens of corresponding diverging power. In the example given a $-2\cdot66$ D, or a $15''$ concave lens, will be the correcting lens, because this will neutralise the $+2\cdot66$ D refractive power which is in excess.

The distant point of distinct vision of the eye in this example when no lens is placed before it will be 15 inches off; or, in other words, the rays of nearest approach to parallel rays which the uncorrected eye is able to focus with accuracy are the rays with that degree of divergency which they have when they start from a point placed at a distance of 15 inches from the eye. A $-2\cdot67$ D lens held in front of the eye causes parallel rays from distant objects to have the same degree of divergency as the rays have which proceed from an illuminated object at 15 inches' distance from the eye. The eye is thus rendered competent to form distinct retinal images of objects at infinite distances just as much as it was able to do of the nearer objects at $15''$ without the lens.

When the precise degree of M. has been ascertained, in correcting it a slight allowance has to be made for the distance at which the trial lens is placed from the eye; and when both eyes together have lenses placed before them, another correction often becomes necessary in practice to compensate for the gain in refraction due to the amount of Acc. in activity which is associated with the convergence of the optic axes when both eyes are employed in regarding an object. Rather weaker glasses are consequently required under the circumstances named than those which the ascertained distant points indicate.

Exaggeration of Myopia from Spasm.—In most cases of seemingly very high degrees of M., it is prudent to try carefully

made on persons who have no knowledge of the effects of lenses, because efforts at deception, if attempted to be practised, are more readily defeated. Other methods of carrying out the investigation are less fully referred to in the text, as the main object is to render this manual as s mple as possible.

Occasional cases will occur, when atropine has not been employed, in which the true degree of myopia may not be exactly found by the use of the $+10''$ lens, because the person under trial has not been able altogether to relax his accommodation when fixing his sight on an object within 10 inches distance; but such instances are rare when the trial is thoroughly conducted. It is also to be taken into account that the action of the convex lens will cause the retinal images of the printed letters to be enlarged, and their magnitude to vary slightly according to the distances at which the letters may be placed in front of the lens. The alterations in the apparent size of the letters, when the lens is kept fixed in position, are not, however, of the same importance as they would be if the distance of the lens were changed instead of the distance of the print, and practically the variations in magnitude which occur do not influence the results to any appreciable extent. This is proved by experiment and by the fact that in cases of uncomplicated myopia and hypermetropia the degree of myopia thus found is shown to be very closely the true one by the corresponding concave lens correcting the vision for distant objects, while the degree of manifest hypermetropia, when atropine has not been used, ascertained by the same means, is equally corrected by the corresponding convex lens.

whether the M. is as high as it appears to be. Counterfeit M., a condition in which no true M. exists, has been already referred to. From causes similar to those which induce counterfeit M., the real condition of M. may become exaggerated in its apparent degree. It will be found in occasional cases that the measure of distance of the remote point of distinct vision is shortened by the spasmodic exercise of a certain amount of Acc., so that the M. is caused to appear higher in degree than it really is. Some myopes read with the print nearer to the eyes than their M. renders necessary. It may be owing to faulty posture, to a habit of reading in a bad light, of reading badly printed books in small type, or to other causes. But under the condition named, an unnecessary amount of Acc. is exerted, and this accommodatory exertion is liable to become so habitual that it cannot be easily relaxed whenever the eyes are similarly employed. The employment of atropine will suffice for the detection of the exaggeration. It prevents the exercise of Acc., and by thus removing the added refractive power will show, in any such case of exaggeration, that the M. is not so high as the tests without atropinisation made it appear to be. If, in such a case, the apparent, but exaggerated, amount of M. were fully corrected, the true amount of M. would be over-corrected; and constant exercise of Acc. would be required to neutralise this excess of correction even for vision at the farthest distance. The correcting glass, owing to its power being too high, would render the eye in question practically hypermetropic.

Over-correction of M.—If the use of the lenses which have been ordered for correcting the M. be found to be all that is desired as regards distant vision, but is attended with discomfort, and aching when the eyes are employed in work at a moderate distance from the eye, leading to the inference that the Acc is not effective for work at the distance indicated as would be the case if the eye were Emc., it should be ascertained if the M. has not been over-corrected by the lenses supplied, and the eyes brought into a condition of H., as described in the previous paragraph. The eyes should be again tested after they have been subjected to the influence of atropine. If there be no over-correction, but the concave lenses ordered are found to be only equivalent to the excess of refraction, or, in other words, to the degree of M. which they have been calculated to neutralise, it becomes evident that the Acc. from some cause or other is alone at fault, and cannot be exerted to the extent necessary for obtaining clear vision of near objects when lenses fully correcting the M. are placed before the eyes. It is best under such circumstances to prescribe two sets of glasses—one set for fully correcting the M., but only to be employed for distant vision, a second set for use at a relatively near distance. The distance for which the second set should be calculated may vary according to the nature and position of the work to be done with their aid—for

such work as reading, for seeing notes of music, or for some mechanical occupation. But under no circumstauces should the glasses
be calculated for a distance very close to the eyes, or with their use
all the discomfort and difficulties attending undue convergence and
strain of Acc. will still have to be encountered. A reading distance
of ten inches or a foot from the eyes should at least be provided for.
This is especially necessary when the degree of M. is so great as to
place the distant point of distinct vision very near to the eyes. It
is important to remove this point further off and so prevent the
fatigue and nerve irritation entailed by accommodatory strain and
excessive convergence.

There is another reason why glasses for full correction should
not be worn at near work in high degrees of M. The higher
degrees of M. are generally associated with some amount of Amblyopia, and as the retinal images are considerably lessened in size
by concave glasses of high power, this diminution, together with
the co-existing amblyopia, impel the myope to bring the objects still
nearer to the eyes. An additional strain is thus caused on the
relations between the Acc. and convergence, asthenopia results,
and if posterior staphyloma exists a tendency to its increase is pro-
moted. The calculation must be made by lessening the dispersive
power of the lens required for full correction to the extent of the
power of the lens representing the Acc. that must be exerted to
give distinct vision at the distance to which the lens is to be
adapted. Thus if a concave lens of -4 D will correct vision
for parallel rays, a lens of -4 D $+.2\cdot75$ D ($2\cdot75$ dioptrics being
nearly the equivalent of 15 inches), or a lens of $-1\cdot25$ dioptrics,
or of 32-inch focal length, will suffice to correct vision for a distance
of 15 inches.

It is sometimes said that in M. of a pronounced degree, the
ciliary muscle becomes wasted from disuse, and Acc. cannot be
exerted. But this is not so in most ordinary cases of M. of high
degree, for when the range of Acc. is tested, it is generally found
to agree with the range proper for the age. Yet in such cases, if
the M. be fully corrected, the exertion of continued exercise of V.
at relatively near distances while the glasses are before the eyes
will be attended with difficulty, and will be followed by ocular aching and weariness. In these cases no solid foundation exists for
attributing the symptoms to atrophy of the ciliary muscle; the
difficulty seems to arise from the fact that during the period no
relief was afforded to the myopic condition, the Acc. was associated
with strong convergence, the range over which the Acc. was
exerted was a very limited one near the eyes, and thus, from
acquired habit and adaptation, it is hard to exercise the Acc. over
a range that is more distant. If the M. be $=4$ D, and this be
wholly corrected, it might be anticipated the eye would be placed
in the same condition as an emmetropic eye, and the Acc. be set

free to exert itself from within infinite distance to the natural near point for the age, without inconvenience, but practical experience shows it is not so. Under such circumstances it is better to fix the distance at which work is to be done, and to correct the excess of refraction sufficiently for obtaining clear images of objects at that distance. The patient with M. = 4 D wishes to read music at a distance of 20 inches, then 4 D − 2 D = 2 D will represent the excess of refractive power which will have to be neutralised, and a − 2 D lens will accomplish the object. The person will then be in the condition of M. = 2 D, instead of M. = 4 D, and the distant point of distinct V. will be at 20 inches. If the individual be 20 years of age, with 10 D of Acc., a range of clear V. will be afforded from R at 20 inches' distance to P. at about 3⅓ inches from the eyes, and the transference of the range of Acc., which, before partial correction of the M., was from nearly 3 inches to 10 inches from the eye, to the range from 3⅓ inches to 20 inches will be accomplished without any trouble. Moreover the notes and letters will appear more nearly of their natural size with the 2 D glasses than they would if concave glasses of 4 D were worn, and this adds to the ease of seeing at the distance calculated for. If the person carries folding glasses of − 2 D, when he looks through them doubled together he will be able to see clearly the most distant objects exposed to view with a single eye, or if he places them in front of the − 2 D spectacles distant objects will be rendered equally clear to both eyes.

Extension of Reading Distance in High Degrees of Myopia.—In like manner in a case where the M. is very high in degree, as 10 D for instance, when it would be important to remove the reading or working point of distinct vision from a distance so near to the eye as 4 inches to some given distance farther off, the same proceeding may be followed as in the last example given.

Supposing it be desired to remove it to a distance of 12 inches. Then the M. being = 10 D, and the distant point of distinct vision of the eye 4 inches, − 10 D would correct the eye for parallel rays or distant vision, and (− 10 D + 3·33 D), or − 6·66 D, would suffice to remove the distant point of distinct vision to a distance of 12 inches from the eye.

Tests in Military Practice for High Degrees of Myopia.—Before concluding the remarks on myopia it may be well to refer to the plan of establishing the presence of high degrees of myopia by means of concave lenses. It has already been mentioned that formerly the army optical and ophthalmoscopic case included spectacles fitted with − 6″ lenses for this purpose. They were originally introduced for the ready detection of degrees of M. = $\frac{1}{12}$, 3·33 D, and upwards. They had been used in the Austrian Army at the suggestion of Stellwag von Carion for determining, when high degrees of M. were urged by conscripts as a plea for exemption from military service, whether such degrees of M. did really exist, or otherwise ; the degree of M. being deduced from the *near* point at which small print could be read when the $-6''$ lenses were worn. The rule laid down was that if a man, when wearing $-6''$ spectacles could read small print (No. 2 Jäger) at or within $6''$ distance from the eye, he was obviously so myopic as to be unfit for military service, for he would be myopic $\frac{1}{12}$ or upwards. Although the conditions of military recruiting differ greatly in foreign armies from the conditions in the British army—M. in an exaggerated degree being one of the defects most frequently alleged as a cause for exemption from service, while, on the other hand, what British surgeons have usually to guard against is its attempted concealment or depreciation in degree where M. exists—still, if the results of so simple a rule were to be relied upon, its application would be capable of being turned to advantageous account in the British, no less than in continental armies, though employed for different ends.

A myopic eye of $\frac{1}{12}''$, $+3\cdot33$ D, is converted into the condition of a hypermetropic eye of $\frac{1}{12}''$, or $-3\cdot33$ D, by $-6''$, or $-6\cdot66$ D, lenses being placed before it. A lens of $-3\cdot33$ D would suffice to neutralise the M. ; a lens, therefore, of $6\cdot66$ D will not only neutralise it, but will leave a refractive deficiency of $3\cdot33$ D. At the age of the recruit, 18 to 25, the power of accommodation is about $=\frac{1}{4}$ or 10 D. Of this, $3\cdot33$ D would be required to neutralise the artificially produced $H = -3\cdot33$ D, and there would remain dynamic refraction $= 6\cdot66$ D, and this when fully exercised would bring the near point of distinct V. to a distance of $6''$ from the eye. Eyes with higher degrees of M. than $\frac{1}{12}$ will be able to obtain still nearer points when the $-6''$ lens is before them. If the amount of M. were equal to $6\cdot66$ D, the $-6''$ spectacles would just neutralise the M., and with the full exercise of Acc. $= 10$ D, the near point of distinct V. would be brought to 4 inches' distance from the eye. Theoretically, therefore, the test of reading within $6''$ with a dispersing lens of $-\frac{1}{6}''$ power should prove the existence of M. $=\frac{1}{12}$ or above, and would exclude M. less than $\frac{1}{12}$, or $3\cdot33$ D. But, practically, myopes with M. no higher than $2\cdot75$ D, or $2\cdot50$ D, by practice in the exercise of Acc. can read within $6''$ with the $-6''$ lens, and their employment for deciding the presence of such high degrees of M. as $3\cdot33$ D or upwards, not being found reliable, was abandoned. It was proved in one instance in Belgium that instruction in the means of fraudulently escaping conscription by practice in the use of concave lenses of the official limit, or, in other words, in developing dynamic Acc., was systematically given as a matter of trade.

The rule in the Austrian Military Medical Service, by the more recent recruiting regulations of 1883, as regards the disqualification of M. for engagement in the army, is the following. A myopic recruit, with a far point of 12 inches or less, if he is able to read printed letters, or to recognise other characters, of $\frac{1}{3}$rd of a Vienna line in height and corresponding breadth at any distance from the eye with concave 4-inch spectacles, is to be rejected as totally unfit for military service without any further examination. The $-4''$, or -10 D, spectacles are to be close to the eye, care is to be taken that the recruit really looks through them, not under them, and a good light is to fall upon the print.

The conditions described afford a proof that the degree of M. of the person under observation is not less than $\frac{1}{12}$, or $3\cdot33$ D, though it may be higher. Supposing the M $= 3\cdot33$ D, there would be Acc. at the age of the conscript about 10 D, so that the excess of refraction above Emmetropia and the dynamic refraction of Acc. together would be equal to $13\cdot33$ D. He could obtain a near point for reading at 3 inches' distance from the eye without the glass. If then a -10 D lens be applied to the eye, the Acc. will be neutralised, and there will remain $+3\cdot33$ D, or M.$=3\cdot33$ D, or $\frac{1}{12}''$. Suppose the far point distance of distinct vision without a glass appeared to be 6 inches from the eye, or, in other words, the M. was alleged to be $\frac{1}{6}$, or $6\cdot66$ D. The mode of proceeding if practised would equally afford a decided proof of its truth if the statement were correct. For a M.$=6\cdot66$ D with Acc.$=10$ D would together amount to $16\cdot66$ D, and $16\cdot66$ D-10 D would leave an excess of refraction, or M.$=\frac{1}{6}$, or $6\cdot66$ D. The concave 4-inch lens neutralises the Acc., and leaves the myopic excess of refraction to act alone. If the eye had been an Emc. eye, and had assumed a distant point of 12 inches, employing for the purpose a portion of its normal Acc., it could no longer read at the distance named with a $-4''$, or -10 D, lens; the eye being Emc., and its full amount of Acc. neutralised by the lens, it would only be capable of adjusting parallel rays and seeing distant objects of suitable dimensions.

It is stated in the Austrian Military Instructions for the surgical examination of conscripts (1883) that ' Many years' experience and very numerous trials have proved that a myope can only satisfy this test if his myopia is above $\frac{1}{12}''$, which is laid down as the limit of fitness for military service.'

Influence of Myopia on Military Service.—For an account of the practical effects of different degrees of M. in respect to the requirements of military service, see Chapter IX. The regulations under which the faulty vision resulting from M. is ordered to cause the rejection of men seeking enlistment as recruits are also given in the same chapter.

HYPERMETROPIA.

Definition.—Over-measure as regards the distance of the principal focus of the dioptric media of the eye in relation to the measure of the optic axis.—The refractive power of the eye, when in a state of repose as regards accommodation, is below what is necessary for forming clear images of objects upon its retina. Parallel rays from distant objects are not brought to a focus by the time they reach the retina, but would unite in a focus, if they were not stopped in their course, at a point beyond the retina; or, in other words, the retina is in advance of the principal focus of the eye. Only rays reaching the eye as convergent rays can be focussed on the retina (see fig. 47).

Optical Conditions.—In H. the antero-posterior diameter of the eye is, as a rule, disproportionately short, and hence its deficiency of refractive power. Similar optical effects would result from the refracting qualities of the dioptric media being too low in relation to the length of the antero-posterior diameter, but the short diameter is the condition usually met with.

Causes.—H. is often congenital. It is not unfrequently due to hereditary conformation of the eye, and, like myopia, is often found to exist in several members of the same family. While the myope's eye is usually a full and prominent eye, the eye of the hypermetrope is commonly a shallow and flat eye. In cases where the hypermetropic state is found to be excessive, the eye appears as if it had been stunted in growth : it is diminutive in form and short in all its dimensions.

Similar optical conditions to those which characterise H. may be induced by any circumstances that lead to flattening of the globe of the eye, or of one or more of its component structures ; as removal of the crystalline lens by displacement or by operation; or sinking of the cornea from any cause. It sometimes shows itself as age advances, becoming noticeable after the patient has arrived at full manhood, without any apparent cause beyond the natural changes due to increased years. It is then associated with *presbyopia*. This form of H. has been designated by Donders *acquired* H., to distinguish it from *original* H., due to early ocular conformation. (See remarks under Emmetropia, p. 59.)

Farthest Point of Vision.—This is sometimes spoken of as *negative*, because the hypermetropic eye has no objective distant point of distinct vision. Only convergent rays can be brought to a focus on the retina, and such rays do not proceed from any natural objects—they can only be produced artificially, as by a converging lens. The Hc. eye has only a virtual distant point, and this is situated at the distance of the focal point to which parallel rays would be produced after traversing the dioptric media if they could pass beyond the retina. In order that parallel rays may be focussed on the retina itself of the Hc. eye, they must be first altered in direction ; they must have such a convergence given to them before entering the eye as would cause them, supposing their course to be unaltered in direction by the action of any refracting media, to meet at the distance of the focus above named. The convex lens that would give this amount of convergency to the parallel rays proceeding from objects at infinite distance, makes up for, and will serve to represent, the deficient converging power in the eye itself, or, in other words, the amount or *degree of hypermetropia.* Hence the remote point of distinct vision of the hypermetropic eye has sometimes been described as being at a distance *beyond infinity* equal to the value of the negative lens representing the deficient convergent power which the addition of the positive lens supplies. This of course is a mere optical expression, but its employment is convenient when it is desired to signify the different positions of the remote points in different degrees of H.

Thus, if a + 2 D lens supplies the necessary amount of converging direction to parallel rays that will enable them, in addition to the refracting power of the eye itself, to be brought to a focus on the retina of the hypermetropic eye, the remote point of vision of this eye may be stated to be − 2 D beyond infinity.

Nearest Point of Vision.—This varies in position according to the degree of H., but is always farther off from the eye than it is in emmetropic persons at corresponding ages of life.

Symptoms.—When a person affected with well-pronounced H. looks intently for a short time at small objects, as in reading and writing, the letters become blurred and seem to run into each other. The vision of distant objects is more limited in range than normal, though the patient himself often fancies he can see well at a distance. The hypermetropic marksman, although he only uses one eye in aiming, cannot adjust vision for distant objects, as in trying to hit the bull's-eye of the target at long-range rifle practice, without an amount of accommodatory or muscular strain proportionate to the degree of H. Still less can he do so for near objects, such for example as the ' back and fine sights ' of the rifle. As part of the natural amount of accommodation possessed by the hypermetrope has to be employed for getting a less indistinct view of the distant bull's-eye, a less amount remains for use in the effort to rapidly adapt vision to the distance at which the nearer objects are placed. When both eyes are habitually employed in looking at near objects, especially such as call for close and accurate observation, a sense of ocular weakness results, and fatigue and aching are quickly produced; the patient suffers from symptoms of *asthenopia* (see Asthenopia). This ocular strain is rendered excessive, and the results of it are much more severe and occur more speedily, when a Hc. soldier is not only engaged in occupations requiring intent and anxious observation, but when his eyes at the same time are irritated by subjection to the heat and dazzling glare of a tropical sun. Examples of this occurred among Hc. men engaged on active service in Egypt and Burmah. The symptoms of asthenopia are the more marked in proportion as the degree of H. is greater, when much anisometria exists, and are especially so if the general health becomes deranged and the subject becomes debilitated in consequence.

Association of H. with Strabismus.—H. is not unfrequently associated with convergent strabismus, and this circumstance seems to arise in the following way. The hypermetropic eye being only able to bring convergent rays to a focus on the retina without the exercise of accommodation, and no such rays existing naturally, the accommodatory apparatus is subjected to a constant exertion, in order to obtain more convexity of the crystalline lens, and thus to make up for the hypermetropic deficiency, and to lessen the diffusion and indistinctness of the retinal images of external objects, even the most distant. This exertion is increased in proportion as the rays entering the eye are rendered more divergent by objects being brought close to the face, as occurs in reading, and the strain is of course all the greater when the work at near objects is prolonged and frequently repeated. Hence the frequent appearance of strabismus in hypermetropic children at the ages when they begin to learn to read and write. The hypermetropic child, generally a child of early age, unconsciously tries to'lessen the excess of strain on the ciliary muscle by contracting the internal recti muscles in order to obtain the advantage of the increased accommodation for near objects which is derived from the association of the accommodatory effort with convergence of the optic axes. The normal balance of power between the internal and external straight muscles, and, consequently, the evenness of their concerted influence over the position of the eye to which they respectively belong disappears; the internal straight muscles from constant exertion either become more developed and stronger, or, from their habitual state of contraction, lose some of their subjectiveness to extension; and a state of constant convergence results, even when the eyes are at rest. Under these conditions such co-ordination of the accommodation and direction of the visual lines as would be necessary for clear binocular vision of objects relatively distant ceases to be attainable, and parallelism of the visual lines for objects at infinite distance becomes impossible. With the intersection point of the visual lines under this convergence near to the eyes the images of objects farther off will not fall upon corresponding points of the two retinæ, and the inconvenience and confusion of double vision will result. Under these difficulties the young hypermetrope finds it less irksome to get a comparatively clear and easily obtained monocular image than to make repeated efforts to get binocular vision which is attended with so much trouble. One eye is turned further inwards by its internal rectus muscle, at the same time that the other eye is turned sufficiently outwards by its external rectus muscle to bring the image of a more distant object on its macula lutea by which a distinct view of it is obtained; a decided squint results : and the image of the object in the con-

·verged or deviating eye falling upon a less sensitive part of the retina, is ignored mentally, or, in other words, ceases to exist so far as the sensorial part of the act of seeing is concerned. For a time monocular vision is carried on equally by each eye in turn according as the position of an object to the right or left of the face may dictate, but at last vision devolves principally on one eye while the other is almost constantly turned inwards. This change takes place more readily when one eye is naturally weaker than the other in respect to retinal power, or has a higher degree of H. than the other eye. These abnormal relations of the recti muscles at last become constant, and a fixed convergent strabismus is established. When strabismus has thus been acquired, its tendency to become permanent is increased by the fact that the deviating eye gradually becomes retinally still weaker, and loses sensibility from disuse. It is for the purpose of neutralising the ill effects of continued strain upon the accommodatory apparatus in the efforts to obtain accurate vision of near objects, that convex glasses are recommended to be constantly worn for the treatment of H.; and it is by these glasses neutralising the H. that the strabismus so often induced by it may be prevented when it is properly treated on being first noticed.

Manifest and Latent Hypermetropia.—As the Hypermetropic eye is not able even to bring rays from distant objects to a focus on the retina by its unaided refractive power, it supplements it by using some of its accommodatory power in order to obtain clear retinal images of the objects looked at. Still more powerfully has it to exert its accommodation in order to see near objects clearly. Thus the ciliary muscle is kept in a constant condition of exercise and tension. This habitual association of the act of accommodation with the act of vision at all distances leads to the loss of voluntary power of completely separating one from the other. In early life when all the structures of the eye are very tractable, and there is plenty of Acc. to be spared, the whole of the H. may be concealed by the Acc. supplied, but as years advance, and there is less Acc. available, and a continual struggle to use as much as possible of this Acc. in association with convergence for obtaining distinct vision at a nearer distance, the H. becomes deprived of a portion of the help it had previously derived from the Acc., and the H. becomes more and more manifest. At last late in life, there is no Acc. to be lent to it, and then the total amount of H. is rendered manifest. But until this period arrives the full amount of H. is not shown unless all power of accommodation is artificially removed. This can be done by producing complete ciliary paralysis through the agency of atropia. The power of Acc. being thus removed, that p rtion of the deficiency of refractive power which was sup-planted, and so concealed by its agency, is brought under observa-tion. Hypermetropia, therefore, of ordinary degrees usually consists of a certain amount of deficiency of refractive power which may be rendered apparent while Acc. is exerted; and of another amount which, naturally concealed by Acc., becomes apparent only when Acc. is artificially prevented or has disappeared from age. The former deficiency is known as Hm., *manifest hypermetropia*; the latter as Hl., *latent hypermetropia*. Obviously H., or the total amount of refractive deficiency, is composed of Hm. + Hl. If the degree of H. be moderate, or the hypermetrope be very young, it may be entirely latent—that is, only apparent after paralysis of accommodation. H. may therefore sometimes exist without attracting attention or being at once readily diagnosed. Usually, however, though slight in degree, and not noticeable at first, it becomes so increased by fatigue, especially as age advances from continued occupation at near objects, or by muscular weakness when the general health is impaired, that the symptoms of H. become apparent, and the diagnosis of it is rendered sufficiently easy.

Subdivisions of Hypermetropia.—A further division of H. has been made by Professor Donders into (1) *absolute* H., in which the rays from distant objects are not able to be focussed on the retina, but their focus still lies behind it, even with the aid of full power of Acc. and the strongest convergence of the optic axes ; *relative* H.. in which the rays from distant objects can be brought to a focus on the retina by the exercise of Acc. and convergence of the optic axes combined ; (3) *facultative* H., in which the rays from distant objects can be brought to a focus on the retina with parallel optic axes, either with or without convex glasses.

Diagnosis of H. may be effected subjectively by the following modes of observation : namely, (1) by external signs, (2) by test-types, (3) by lenses, (4) by + 4 D spectacles, (5) by correction, and (6) objectively by the ophthalmoscope.

1. *By External Signs.*—The eye, as before mentioned, frequently has a general flat appearance and seems smaller than normal, as if it had been stunted in its growth, so that, in consequence, the space beneath and between the eyelids is not filled out, as it is in a fully grown eye. Sometimes the iris may be observed to reach nearer to the cornea than usual, owing to want of depth in the anterior chamber of the eye, while the pupil is inactive and relatively small. An abnormal hollowness of the space between the eyeball and outer canthus may also be generally noticed on drawing the orbicular coverings aside. The appearances just described are, however, occasionally absent, or only very slightly indicated in eyes affected with Hc. vision.

2. *By Types.*—If the H. exist in high degree, Snellen's types cannot be read by the subject of it clearly at their regular distances without the aid of lenses ; but if it be moderate, the larger and more distant type may be read for a time, but with more or less difficulty. This is especially the case with young Hc. persons who have an abundance of accommodatory power, and who

can exercise a portion of it with comparative ease in order to
bring the parallel rays from distant objects to a focus on the retina.
Such young persons may be able to read Sn. No. 6m. type at a
distance of twenty feet without difficulty, but, on placing weak
convex lenses before their eyes, the type will be seen with equal
clearness.   The power of the + lenses which can be thus employed
without lessening the distinctness of view of the types at the
distance named will vary in proportion to the amount of Hm. of
the person under examination.   With older persons, when the
types and the distance named above them are maintained, the eye
soon becomes fatigued and the letters indistinct.   In trying to
read, a tendency for the optic axes to converge may sometimes be
noticed, and the patient will often be observed to squeeze the eyelids together for the purpose of contracting the interpalpebral
fissure.   The association of the internal recti muscles, with the
exercise of accommodation, explains the occasional converging aspect
of the eyes in the efforts at reading.   The reason why the larger
and more distant types are perceived with less difficulty than the
smaller and nearer type, is simply due to the fact that the accommodatory strain is increased in proportion to the proximity of objects,
owing to the rays that enter the eye from near objects being more
divergent in proportion to their nearness; and the difficulty is
greater in proportion as age is greater, because the power of
Acc. diminishes as years advance.

When types of smaller sizes are used, the Hc. person, if he be
of the ordinary age of a recruit or soldier, will be generally
observed to carry them nearer to his eye than the regulated distances in trying to read them.   He will be able to distinguish
them better when close to his eye than at a distance away from it.
The patient does so because, with the greater convergence of the
optic axes when the object is close, he is better able to exert his
accommodatory power to assist in counteracting the difficulties of
reading, which are due to his hypermetropic condition of vision ;
and, in addition, because the retinal images of the letters are increased in size, and therefore more readily perceptible, while the
circles of dispersion are not enlarged in a corresponding proportion,
owing to the pupillary aperture becoming smaller on the near
approach of the letters to the eyes.

From this circumstance, in diagnosing by types, H., when excessive, may be mistaken for M. combined with amblyopia.
Distant objects are seen indistinctly—type of moderate size has to
be held close to the eye to be read—and very small type cannot be
read at all, as happens in M. complicated with Ambl.   The
diagnosis may be established by the fact that in H. distant objects
are seen more distinctly, and the moderately sized types can be read
further off, with the aid of convex glasses, while the same glasses
would produce exactly opposite results in M.   But even without

+ lenses the diagnosis may be established, for it may be observed that in H. the larger Sn. types may be read quite as well, if not better, as regards *relative* distance, than the smaller types, which would not be the case with M.

3. *By Lenses.*—This, or the Donders, method of diagnosis is carried out in the same way as explained in the diagnosis of M. In the case of H. the —lens is found to make V. worse, while the weak + lens either makes V. clearer, or sensibly relieves it of a certain amount of strain, while V. remains as clear as before the application of the glass. Convex lenses of gradually increasing power are then applied, and the strongest glass that can be borne at the same time that the types are seen clearly at the normal distance gives the measure of the degree of H.

4. *By the* +4 D *Spectacles.*—When the convex 10″ lenses are worn, the Hc. eye is able to read type of moderate size at a distance beyond 10″ from the eye without the use of atropine. The distance at which the type can be read will be increased when the power of Acc. has been previously taken away by paralysing the ciliary muscle with atropine.

5. *By Correction.*—When the true degree of H. is ascertained, if no complication exist, the proper convex lens for that degree will completely correct the abnormal condition. It supplies the refractive power, which is missing in the eye itself, and produces the same effect as if the deficiency did not exist.

6. *By the Ophthalmoscope* (see p. 107).

Expression of the Degree of H.—The degree of H. may be either expressed by the power of the convex lens which would so act upon parallel rays as to give them the amount of convergence that would cause them to meet at a similar distance to that of the virtual remote point of the hypermetropic eye, for rays having such a convergent direction when entering the eye would be caused to meet upon the retina; or it may be expressed by the power of the concave lens which represents the amount of deficiency in refracting power of the eye relatively to emmetropia. The latter mode is the most convenient and simple. The H. will then be expressed by the amount of the defect, and not by that of the lens which corrects it. H. $= 2$ D then signifies that the deficiency of refracting power in the eye is equivalent to and represented by a $- 2$ D lens, and that this deficiency will be neutralised or corrected by a $+ 2$ D lens.

To determine H. or the Degree of Hypermetropia; or, in other words, to ascertain the total deficiency of refracting power as compared with emmetropia, by means of the $+ 10″$ or $+ 4$ D spectacles.

Having thoroughly paralysed the power of accommodation by the use of a strong solution of atropia, and noted the distant point of distinct vision of each hypermetropic eye, examined singly, with the $+ 4$ D or $+ 10''$ lens before it, deduct from the power of this lens the inverted value of the distant point, and the difference will give the degree of H. in the eye examined.

Example.—Suppose the distance point is found to be 15'', then H. $= \frac{1}{10} - \frac{1}{15} = \frac{1}{30}$.

Explanation.—Let x = the refracting power of the eye under examination; let a = the refracting power of an emmetropic eye; $\frac{1}{10}$ = the power of the lens to which the eye has been subjected; and $\frac{1}{15}$ = the power of the lens which would give the ascertained distant point of 15'' if the eye were emmetropic. The deficiency of the refracting power in the example given is therefore equivalent to the difference between a $\frac{1}{10}$ and a $\frac{1}{15}$ lens. This deficiency can only be in the eye itself, and its refracting power must obviously be increased to a corresponding degree to bring it to a par with an emmetropic eye.

Therefore $x + (\frac{1}{10} - \frac{1}{15}) = a$; or $x + \frac{1}{30} = a$; or $x = a - \frac{1}{30}$. H., the total deficiency of refracting power, is $= \frac{1}{30}$.

Or, converting the distances measured by inches into dioptrics, as explained in calculating the degree of myopia, there result: $2·66$ D $- 4$ D $= - 1·34$ D, the amount of deficiency of refractive power, or degree of H.

If the distant point of distinct vision with the $+ 4$ D lens be infinite, then H. $= \frac{1}{10} - \frac{1}{\infty}$, or $= \frac{1}{10}$ or $- 4$ D.

If H. be suspected to be in excess of this amount, a stronger lens than a $+ 4$ D lens will be necessary to ascertain its degree by this method. The $+ 2''$ lens employed with the ophthalmoscope is generally available to army medical officers for the purpose. If H. $= \frac{1}{6}$, the distance point of distinct vision with the $+ 2''$ lens will be at $3''$ for H. $= \frac{1}{2} - \frac{1}{3} = \frac{1}{6}$. This will probably be the highest degree of H. to be met with, unless the lens be absent either from accident or operation (aphakia). But it is absolutely necessary that only one eye be examined at a time, and that it should be atropinised, for the spasm of the ciliary muscle associated with the convergence required for so short a range would otherwise prevent a proper diagnosis being arrived at.

If it be only required to determine Hm., or the degree of manifest hypermetropia, in all moderate cases the same *modus operandi* as above described may be followed, and the previous use of atropia may be dispensed with. But each eye should still be examined singly.

To find the Correcting Lens.—The total deficiency of refractive power having been determined, it may be corrected by a lens supplying the amount of converging power which is deficient, providing no accommodatory power is exerted at the same time. In the first example given, a $+ \frac{1}{30}$ or approximately a $+ 1·25$ D lens will be the correcting lens, because it will supply the refractive power which has been proved to be absent.

The absence of this amount of refractive power was shown in the experiment already above explained. The action of the lens will be such that when the patient looks at distant objects, the parallel rays proceeding from them, in passing through the lens, will be caused to assume a converging direction before they fall on the eye; and the angle of convergence given to the rays by the action of the correcting lens will exactly correspond with that degree of convergence which is wanting owing to the hypermetropic formation of the eye itself. The parallel rays from distant objects will thus become focussed on the retina, and clear vision obtained without exercise of accommodation.

This is the general principle on which correction of H. is effected; but in practice several modifying circumstances have to be specially considered. The hypermetrope, when he is not under the influence of atropia, cannot fully give up the Acc. which he has been accustomed to use constantly as a substitute for the deficient refractive quality of his eyes. Even when convex glasses are placed before his eyes, exercise of part of the Acc. to which they have been accustomed is still continued. If, therefore, while this amount of Acc. is still active, convex glasses equivalent to a full extent of the H. be furnished, the eyes will be put into the condition of myopic eyes, with an excess of refracting power for rays from distant objects. The recommendation of Professor Donders for the correction of H. was that glasses should at first be supplied equivalent in power to the total manifest H., together with one-fourth of the latent H. After a certain interval, if the glasses be constantly employed, the efforts of Acc. will become partially relaxed, but in the course of time the use of the remaining Acc. will probably again induce symptoms of asthenopia. The glasses at first given can then be strengthened to the amount of another instalment according to circumstances of the Hl., and, should the symptoms return after a further period of time has elapsed, the glasses may then be strengthened to the full amount of the H. The final adjustment, if it be satisfactorily borne, will place the subject of the H. in the same condition as an emmetropic person of corresponding age. The full amount of accommodatory power possessed by him will be rendered available for its normal purposes.

Influence of H. on Military Service.—The conditions under which H. disqualifies for military service in the British Army are explained in Chapter IX., in which the authorised modes of conducting the visual examination of recruits and soldiers are explained. On all occasions in which spectacles are allowed to be worn by soldiers, as in range practice, and at the School of Musketry, the use of correcting lenses by hypermetropes, even though the H. may be very low in degree, is of great advantage. It takes off the strain of the Acc. which is always present when a Hc. eye is try-

ing to see distant objects clearly without such assistance, while it
sets free the Acc. for use at near objects, so that at target practice,
especially when the men are young, both far and near objects, such
as the objects painted on the target, together with the fore and
back sights of the rifle, are all seen more plainly and with less
visual exertion and fatigue. The indistinctness of vision that H.
entails, the ocular troubles that accompany it particularly when
the subjects of it are exposed to the glare of tropical light or are
reduced by privation and over-fatigue such as men so often have to
endure when they are on active service in the field, and the constant increase of visual difficulty with increasing years, all these
render men with Hc. vision particularly incompetent for military
service, especially in an army in which correcting glasses are not
permitted to be worn.

ASTIGMATISM.

Definition.—A term signifying that the rays proceeding from a
single point are not, after refraction, reunited in a single point.
Applied to vision, it signifies a state in which there is inaccuracy
of view from malformation and blurring of retinal images owing to
the fact that although some of the rays proceeding from an object
may be brought to a focus on the retina, other rays proceeding
from the same object are not at the same time similarly focussed
upon its retina.

Optical Conditions.—The refractive quality of the astigmatic eye
is not alike in all its meridians, and, consequently, has no single
focus. The eye may be Emc. in one meridian, while in another it
is Mc. or Hc.; it may be Mc. or Hc. in all its meridians, but the
degrees of M. or H. may be relatively different in them; and
lastly, it may be Mc. in one meridian, while it is Hc. in the other.
Occasionally the refractive quality will vary in one and the same
meridian of the eye to such an extent as seriously to interfere with
clearness of view. The term *regular astigmatism* is applied to that
kind of Ast. which depends upon dissimilar curvatures of different
meridians of the eye; *irregular astigmatism* to that which depends
on unequal curvature in one and the same meridian, or to excessive
spherical aberration of rays.

Causes.—Congenital asymmetry of the anterior segment of the
ocular globe, or of the cornea, of such a kind as to cause a greater
curvature of one meridian compared with that of the intersecting
meridian. The absence of perfect sphericity is sometimes recognisable on observing the cornea by lateral illumination. An abnormal position or unequal curvature or other structural peculiarity of
the crystalline lens has been said also to be a cause of astigmatic
vision.

Normal Astigmatism.--All eyes have ordinarily a higher degree of curvature in a vertical direction, and therefore a shorter focal distance than they have in a horizontal direction, but the difference of curvature in the two directions is usually slight, and is not found practically to interfere with correct V. The condition is strictly one of Ast. of the kind known as Regular Ast. ; but, from the fact of V. not being lessened by it, owing either to the moderate degree in which it exists or to the completeness with which in this moderate degree it is neutralised by accommodatory compensation, is sometimes described as *Normal Astigmatism*. An exaggeration of the difference in curvature and refractive power in the two directions gives rise to the inaccurate vision and disturbing symptoms characteristic of Ast as an abnormal condition or ocular defect.

In normal Ast. the meridian of greater curvature does not precisely correspond with the median vertical line but intersects it at an angle of about 15°, passing from the centre respectively upwards

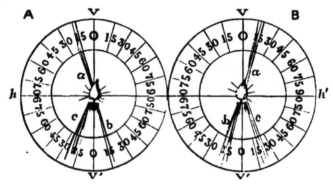

FIG. 48.

A. direction of corneal extensions of rays, *a, b, c*, from a lighted candle as seen by the left eye, B, ditto, as seen by the right eye. (For full explanation see the text.)

and outwards, (*a*), downwards and inwards, (*b*), and again downwards and outwards, (*c*). My colleague Dr. Macdonald has maintained that the difference in direction of curvature in the two eyes of the same individual is in accordance with a bi-lateral symmetry which has apparently escaped notice hitherto. A simple experiment serves to illustrate the fact. When the left eye is closed and a small centre of light, as a lighted candle, that has been placed a few feet in front of the observer, is looked at by the right eye, the head being raised, but the eye turned downward toward the light, a pencil of rays shoots from the light upwards and outwards at an angle a little to the outer side of the vertical; but when the head is lowered slightly, the eye looking in the same direction, the pencil of rays shoots downwards at a similar angle to the inner side of the vertical. When the experiment is repeated with the left eye, the right being closed, the pencil of rays shoots from the light obliquely upwards to the left or outer side of the vertical; when the head is bent forwards the rays shoot obliquely downwards toward the inner side of the vertical. In each instance when the head is maintained upright the angle at which the light radiates is about 15° to the outer or inner side of the perpendicular. The directions of the corneal rays thus produced in the right and left eye respectively are shown in fig. 48, A and B, while the converse corneal curvatures indicated by the directions of the radiation are shown in fig. 49, A and B.

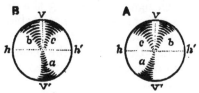

FIG. 49.—NORMAL ASTIGMATISM.

Corneal curvatures indicated by the conditions described in fig. 48. B, right eye ; A, left eye.

This normal Ast. seems to have been designed as an important aid to the perfection of binocular vision of near objects. In demonstration of this view, Dr. Macdonald has shown that when a dot, or a vertical line is placed so far within the focal distance of either eye as to present a blurred aspect, the left side of the dot or line is found to be more clearly defined when the left eye is used, and the right side when regarded by the right eye. In each instance, on the inner side of the object, or that toward the nose of the observer,

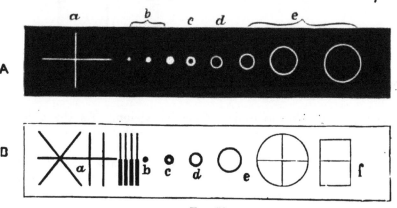

FIG. 50.

A, B, effects of normal astigmatism (see text).

a repetition of more or less indistinct images of the object is perceived, and in this direction a softly shaded effect is produced. Figs. 50, A and B, and 51, C and D, will serve to illustrate the effects just described. When regarding any one of the objects included in them it will be seen, if the two eyes of the observer are emmetropic, that every particular observed by one eye has its counterpart when the object is observed by the other eye. In fig. 50, A,

on closely inspecting the crossed white lines at *a* with either eye
separately, the vertical line will be observed to be brighter and
better defined on the outer side as regards the eye making the examination than on the inner side, where it appears more diffused
and faint. The horizontal line, on the contrary, is more even in
its aspect, its upper and lower borders being similar in appearance.
In fig. 50, B, on experimenting with the crossed black lines, similar
results ensue. In both figs. in observing the dots and circles from
b to *c* corresponding effects will be noticeable. In the smallest
dots, and thinnest lines, the outer side is the more distinct according to the eye observing them, while the black and white centres
in *c, c,* of A and B, are completely obliterated, and in *d, d,* are much
reduced in apparent size.

Fig. 51, C and D, illustrates the combined result of these effects
of normal Ast., in perfecting the definition and obviating confusion

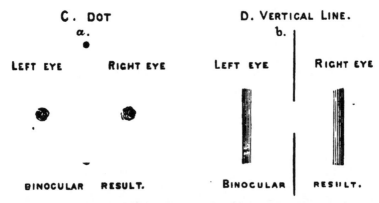

FIG. 51.—COMBINED EFFECT OF NORMAL ASTIGMATISM IN BINOCULAR VISION.

in the aspect of objects under binocular vision. As the objects
are removed from the two eyes acting together, the outer, more
defined, lines or margins which are seen by the eyes separately
approach one another, and at last coalesce; and thus, when they are
brought by Acc. to their true retinal foci, each object is seen singly
and equally defined and distinct on both borders or on all sides.

Symptoms.— All objects, as well those which are near, as those
which are distant, are more or less dulled in appearance, and have
their shapes more or less modified in certain particulars, when they
are regarded by subjects of astigmatism. The change in the shapes
of objects as seen by an astigmatic eye, chiefly concerns their
linear dimensions. The nature of the change can be observed by
placing a cylindrical lens close to an emmetropic eye, and so
practically making it for the time astigmatic. Suppose a $+2$ D
cyl. lens, with its axis vertical, is placed before such an eye, and
an object bounded by straight lines is looked at, the object will

appear elongated in its lateral dimensions, and shortened vertically ;
while, if the glass be turned round, the shape of the object will
alter as it is rotated, and when the axis has become horizontal, it
will appear elongated vertically and shortened laterally.   Thus the
sides of a square object cease to appear equal.   A -2 D cyl. lens
will produce similar results, but the alterations of shape will be in
opposite directions to the alterations effected by the $+2$ D cyl.
lens.   In proportion as the distance of the cyl. lens from the eye
is increased, so necessarily the apparent alteration in shape of the
object looked at through it will be also increased.   If the origin of
the astigmatism be congenital, the symptoms just described will
have been continuous and without much alteration in degree.

Some parts of objects are seen more distinctly than others by
an astigmatic eye.   If the image of an object be sharp and defined
in one direction, it will be rendered indistinct by diffusion of rays
in a contrary direction, or if the image be indistinct from diffusion
in both directions, it will be still further confused by the blurring
being more widely diffused in one than in the other direction.   In
reading, the letters appear badly printed.   Objects presenting
linear intersecting markings, such as patterns with crossed stripes,
are comparatively strongly defined and darker in colour in one
direction, while in the other they appear faintly marked ; or the
whole pattern may become more or less obscure from the diffused
rays from one direction of lines spreading over the lines of the
pattern in the opposite direction.   The astigmatic subject manifestly cannot by any accommodatory efforts bring the rays which
are differently acted upon in passing through the various meridians
of the eye to a focus on the retina at one and the same time, and
he therefore exercises his Acc. to obtain greater clearness of view,
first in one meridian and then in the other, often in rapid succession, so that asthenopia is induced, and adds to the visual trouble
of the patient.   The greater the difference of refraction in different
planes of the eye, in other words the higher the degree of Ast., the
more strongly marked will be the symptoms above mentioned.
The more open the pupil of the eye, the more obvious to the
patient are the effects produced by his astigmatism.   Ast. is often
associated with high degrees of abnormal refraction, both M. and
H., but the ametropic condition with which perhaps it is most frequently associated is H.

Retinal Images of Astigmatic Vision.—In consequence of the
rays passing through the meridian of greatest curvature being
brought to a focus earlier than the rays passing through the
opposite meridian, diffusion of the rays composing the image of
an object is an inevitable result.   The rays passing through the
meridian of greatest curvature will throughout their path towards
their focus be more convergent than those through the meridian of
least curvature, and a section of all these rays on a screen, placed

perpendicularly to the optic axis, would have an elliptical outline with the shortest axis of the ellipse formed in the direction of the most convergent rays. When the more converging rays have been brought to a focus, the rays through the meridian of least refraction will still be in progress, and at this distance a linear image will result with its length in the direction of the rays of least convergence which have not yet been brought to their focus. After the more convergent rays have been brought to a focus, if there be no interruption to their progress, they will cross and pass onwards divergently, while the less convergent rays will be approaching their focus. Here the image will also be elliptical, but the short axis of the ellipse will be in the direction of the rays of least convergence. On these rays arriving at their focus, a linear image

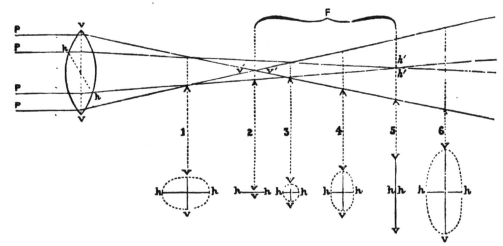

FIG. 52.—DIFFUSION IMAGES IN ASTIGMATISM.

PPPP, para'lel rays from distant object: VV, vertical meridian of refracting medium (astig-matic); AA, horizontal meridian of ditto; 1, 2, 3, 4, 5, 6, transverse sections at distances indicated.

The space (F) between the point of intersection, V'V', and that at A'A', is called the 'focal interval.'

will result, with its length in the direction of the divergent rays which had crossed from the first focus. At a certain point the distances at which the boundaries of the crossed divergent rays, and those of the converging rays still advancing towards their focus, are separated from each other will be alike, and at this point, and this alone, the image on a screen, such as the retina, would be a circular image. The mode of formation, and the varieties of form, of the diffused images just described, are illustrated in fig. 52.

Amount of Astigmatism.—This is determined by estimating the difference in refracting power of the two meridians of the eye in which the inequality of refraction is most marked. Whenever Ast. exists, there must be one meridian in which the refracting force is greatest, and another in which it is least. These two meridians cross at right angles to one another or nearly so. They are the *principal meridians* of the astigmatic eye; the intervening meridians varying in refractive power according as they approach one or other of the two principal meridians. Either the expression of the difference between the refracting power of the two principal meridians, or of the power of the lens which would remove this difference by making the two equal, will therefore represent the amount of Ast. in any particular eye.

Rays of light traversing the principal ocular meridian of relatively greatest curvature, and therefore relatively greatest refractive power, will obviously be brought to a focus at some point anterior to the focus of the rays which have traversed the principal meridian perpendicular to it, or the one which has the least refracting power. The space between the two foci, bracketed under F in fig. 52, is known as the *Focal Interval*. The greater the difference in refractive power of the two principal meridians, the longer will be the *focal interval*; the less the difference in refractive power between these meridians, the less will be the extent of the *focal interval*. The measure of the focal interval, or of the lens which will bring the two foci into exact coincidence, affords another means of expressing the amount of astigmatism.

Estimate of Amount of Astigmatism.—The measure of the amount of Ast., or of the difference between the refractive powers of two opposite meridians of an astigmatic eye, must be calculated differently in different kinds of Ast. If (1) the eye be ametropic in one meridian only, the opposite being emmetropic, the amount of ametropia in the one meridian will express the amount of Ast.; if (2) the eye be ametropic in two opposite meridians, and the ametropia is like in kind—that is, if both meridians are Mc. or both Hc., but differing in degree in each—the *difference* between the ametropia of the two meridians will express the amount of Ast.; if, lastly (3), the eye be ametropic in the two principal meridians but the ametropia is unlike in kind—that is, if one meridian be Mc. and the other Hc.—the *sum* of the two degrees of ametropia must be taken to express the amount of Ast. or difference in refractive powers of the two meridians.

Three Kinds of Astigmatism.—The three varieties of regular Ast. referred to in the preceding paragraph are named: (1) Simple Astigmatism; (2) Compound Astigmatism; and (3) Mixed Astigmatism. In form No. 1 one principal meridian of the eye is emmetropic, while in another and opposite meridian there is either an excess or deficiency of refractive power; in No. 2, in both principal meridians, either an excess or a deficiency exists, but in different amounts; in No. 3 there is excess in one principal meridian, while in the opposite one there is deficiency of refractive power as compared with an emmetropic eye. (See fig. 56 at end of chapter.)

Diagnosis of Regular Astigmatism.   Subjective Methods.

1. *By Lenses.*—Though the eye may be hypermetropic, or myopic, neither + nor − *centric* lenses of any power will correct the existing defective condition or materially lessen the amount of acuteness of vision which is missing.

2. *By Lines.*—When vertical and horizontal lines and rows of separate square dots, as shown in fig. 53, are placed in front of the eye, they are not seen with equal definition. When the vertical lines appear dark and defined, the horizontal lines will not be seen

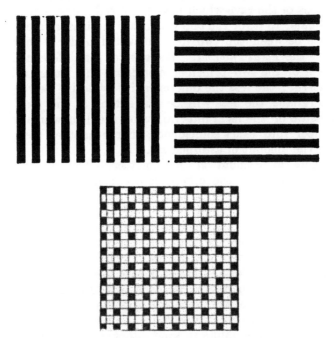

FIG. 53.—VERTICAL AND HORIZONTAL LINES AS TESTS FOR ASTIGMATISM.

with equal clearness, and, *vice versâ*, when the horizontal lines are seen comparatively clearly, the vertical lines will appear hazy. If attentively observed, the indistinct dark lines appear lighter in colour than they are printed, while the white spaces between them appear to be darkened, the general effect being to produce a greyness of colour, as if the black and white lines were mixed together, or as if part of the black were spread over the white, and part of the white spread over the black objects. If the horizontal lines be indistinct at the same time that the upper and lower margins adjoining the white interspaces are undefined, the left and right extremities of the lines will be sharp and distinct. On the contrary, the lines placed vertically which are seen separate and defined, appear very black in colour, without any blending with the white interspaces, but the upper and lower ends of the lines appear to be a little elongated, or present the appearance of a shadow extending beyond each of them.

The lines which are most blurred, and so seen most indistinctly, are those which are in a *contrary direction* to that of the most ametropic meridian of the eye of the observer. Lines, on the other hand, in a direction corresponding with that of the most ametropic meridian, are seen distinctly if one meridian be Emc., or with relatively greater distinctness if both meridians be ametropic. Thus supposing an eye to be Mc. in one meridian but Emc. in the meridian perpendicular to this Mc. meridian, the linear markings of any objects looked at by it will appear hazy and obscure, though defined at their ends in the direction of the Emc. meridian ; distinct, normal in intensity of colour, but elongated at their ends in the direction of the Mc. meridian. Or supposing the whole eye to be Mc., but in different degrees in its principal meridians, lines seen in the direction of the less Mc. meridian will appear to be the most indistinct and confused, those in the direction of the meridian having the highest degree of M. will appear to be the least so. The same will be the relative appearances of objects seen by astigmatic eyes in which the defective refractive quality is Hc. in kind.

If the eye be Mc. in one meridian but Hc. in the opposite one, lines parallel with the meridian in which the defect of refraction is greatest will be seen with most distinctness, while lines parallel with the meridian in which the defect is least will visually appear the most obscure. Thus the preponderating defect as regards form of the astigmatic eye, and the preponderating defect as regards visual effect, are always in opposite directions to each other. It follows, in practice, that the fact of lines being seen in a certain direction more obscurely than in the direction perpendicular to it at once shows the direction in which the astigmatic eye under observation is least ametropic ; and, *vice versâ*, the direction of the lines seen most clearly indicates the direction in which the form of the eye in respect to its refraction is most defective.

The explanation of the visual effects just described is simple. Thus, taking the case in which an eye is Mc. in one meridian and Emc. in the opposite one, the eye being accommodated for the distance at which the vertical and horizontal lines of Snellen are placed, the rays of light emanating from every point in the lines parallel with the Mc. meridian will be brought to a focus in front of the retina, and images of all these points will be confused by circles of diffusion, while the rays falling in the Emc. meridian will be focussed in the plane of the retina and form clear images. Supposing the Mc. meridian to be vertical and the horizontal meridian Emc., the diffusion will of course be in a vertical direction. Vertical lines will therefore appear shadowy and elongated at their upper and lower ends, but there will be no diffusion at their lateral margins, or in the direction corresponding with the horizontal

meridian of the eye.   The diffusion of the rays proceeding from
each dark point upwards and downwards, on the surface of any
vertical line, will be superimposed on the adjoining dark points of
the surface, and the general blackness of the whole line will not be
interfered with.   The only visual defect as regards lines in the
direction of the defective myopic meridian will, therefore, be the
shadowy prolongation of the vertical ends of the lines; the re
mainder of the dark lines, and of the white interspaces, will be
clear and defined.   But as regards lines parallel with the Emc.
meridian, or approaching it in parallelism, the diffusion being still
in the direction of the vertical Mc. meridian of the eye, the upper
and lower boundaries of the horizontal lines, both the black and the
white lines, will become diffused vertically on the retina, while
there will be no diffusion as regards the ends of the lines.   The

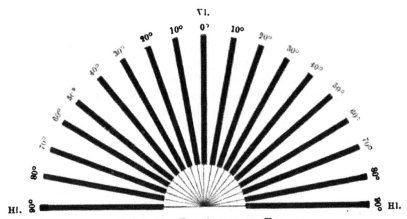

Fig. 54.—The Astigmatic Fan.

vertical diffusion of their images on the retina will thus cause the
series of black and white horizontal lines to appear to be mixed
up together, not only rendering their upper and lower borders undefined, but causing also a general greyness of colour over the whole
series.   The ends of the lines will, however, remain sharp and distinct, because there is no diffusion of their images in the horizontal
direction.   The same explanation is applicable to the cases in which
both principal meridians are ametropic though in different degrees;
the relative visual defects depend equally on the causes just
explained.

3. *By the Astigmatic Fan.*—The same remarks will apply to
lines when they are arranged in the form of the astigmatic fan (see
fig. 54).   The line which appears to be blackest, and to have the
most sharply defined borders, indicates the direction of the chief
ametropic meridian of the astigmatic eye; while the line perpendicular to it will be the most defective in definition, and will
indicate the meridian of the eye which is least ametropic.

If the Astc. fan be extended so as to form a wheel-like circle of spokes or radii, while the central portion is filled up by a series of concentric and equidistant circles, similar effects to those described with the Astc. fan will be manifested, and if the figure be rotated the alternating displacement of the focal adjustment on the retina of different parts of the central circles will produce the effect of a revolving movement in them (see fig. 55).

4. *By a Circular Point of Light.*—When a small round opening is made in a dark screen and light admitted through it, if looked at

FIG. 55. – CIRCULAR EXTENSION OF THE ASTIGMATIC FAN WITH CONCENTRIC AND EQUIDISTANT CIRCLES WITHIN.

by an astigmatic eye at a distance of a couple of feet or so, the cone of light appears elliptical in form, the direction of the ellipse varying as the eye approaches or recedes from the opening. Whatever may be the direction in which at one distance the hole in the screen appears to be elongated, at some other distance from the eye the hole will appear to be elongated in another and generally in a contrary direction.

The variations in the apparent form of the small opening depend on the differences of distance for which the eye is accommodated. If the astigmatic eye is accommodated to the distance at which the small opening is placed, so as to bring the rays of light proceeding from it to a focus on the retina in one meridian, the opposite meridian will · be relatively ametropic, and diffusion of the rays passing through it will result. Thus, taking for example an eye that is Mc. in its vertical meridian and Emc. in the opposite meridian, when the small hole is so placed that the light traversing it is brought to a focus on the retina in the Emc. meridian, diffusion of rays will occur in the direction of the Mc. meridian, and the point of light will appear to be elongated in that direction : the opening will be elongated vertically. If, however, the distance be such that the rays of light are brought to a focus on the retina in the line of the Mc. meridian, the opposite Emc. meridian will practically be rendered for the time ametropic, deficient in refractive power, and diffusion will occur in that direction. If the eye in its horizontal Emc. meridian be accommodated for focussing on the retina rays of light coming from the distance point of the Mc. meridian, the Mc. meridian will have its distant point shortened, and diffusion will occur in that direction. The point of light will appear to be elongated vertically.

5. *By a Stenopœic Hole.*—When the stenopœic opening is placed in the visual line near the centre of the cornea, and the astigmatic eye looks through it at the vertical and horizontal lines of Snellen, they appear as sharp and defined, or nearly so, in all directions, as they do to an Emc. eye.

6. *By a Stenopœic Slit.*—When an astigmatic eye looks through a stenopœic slit at an object, the rays of light proceeding from the object and passing through the slit will be acted upon by the dioptric media of the eye according to the refractive quality of the particular meridian the slit coincides with. If the case be one of simple Mc. or Hc. Ast., and the slit be placed in the direction of the Emc. meridian, both vertical and horizontal lines will appear equally clear. If the slit agree with the ametropic meridian, there will be diffusion of rays in the direction of that meridian, and lines corresponding with the meridian perpendicular to it will be rendered visually dim and obscure. The proper correcting + or − lens, according to the kind of ametropia, will at once remove the confusion. If the slit be applied to any intervening meridian similar effects will occur in a modified degree, according to the position of the meridian. If the case be one of compound Mc. or Hc. Ast , and the slit be placed so as to correspond with either of the principal ocular meridians, the visual effect will be the same as if the case were one of simple Mc. or Hc. Ast. of amount corresponding to the degree of ametropia of the particular meridian to which the slit is applied. If it be a case of mixed Ast., objects will be seen as they would appear if they were regarded by an eye with simple Mc. or Hc. Ast., according to which meridian the fissure is applied to. In all cases a correcting spherical lens will remove the astigmatic

effect produced by the application of the stenopœic slit. It should, however, be remembered that effects similar to those last described would ensue if the stenopœic slit were applied to a myopic or hypermetropic eye of the same degree of M. or H., but without any Ast.

Diagnosis of Astigmatism.   Objective Method.

The existence of Ast. may be determined objectively by the changes of form it produces in the ophthalmoscopic appearance of the optic papilla. When the fundus is observed by the direct mode of examination the optic papilla appears elongated in a direction corresponding with that of the most refractive ocular meridian; when the fundus is observed by the indirect mode of examination, and the image is inverted, the optic papilla appears elongated in the direction corresponding with the least refractive meridian.

Varieties of Regular Astigmatism.—All cases of regular Ast. belong to one or other of six varieties of the three kinds of Ast. before mentioned. They are the following:—

1. *Simple Myopic Ast.* (Am.)—One ocular meridian is Mc., the contrary meridian Emc. Parallel rays passing through the former meridian are brought to a focus in front of the retina; through the latter meridian, in the plane of the retina.

2. *Simple Hypermetropic Ast.* (Ah.)—One meridian is Emc., the contrary meridian Hc. Parallel rays passing through the former meridian are focussed in the plane of the retina; through the latter meridian, have not attained their focus on reaching the retina.

3. *Compound Myopic Ast.* (M. + Am.)—Both principal meridians are Mc., but in different degrees. One meridian presents a maximum of M., the meridian at right angles to it a minimum. Parallel rays traversing the two principal meridians are all brought to a focus in front of the retina, but at different distances in front of it.

4. *Compound Hypermetropic Ast.* (H. + Ah.)—The two principal meridians are Hc., but in different degrees. One meridian presents a greater deficiency of refractive power than the other. Parallel rays traversing each meridian would, if practicable, be brought to a focus at a distance beyond the retina, but the focal distances beyond the retina would be different.

5. *Mixed Ast.; Myopia Predominant* (Amh.)—Both principal meridians are ametropic, but the anomalies of refraction are opposite in kind. M. in one meridian is mixed with H. in the contrary meridian, but the M. predominates.

6. *Mixed Ast.; Hypermetropia Predominant* (Ahm.)—This variety is similar to No. 5, with the exception that H., instead of M., predominates.

The amount of Ast. in each of the foregoing six varieties will vary according to the difference between the maximum and the minimum of refraction in the two principal meridians. Such variations are very numerous. Of the three kinds of Ast. to which the six varieties belong, *i.e.* the simple, compound, and mixed kinds of Ast., the compound is most frequently met with in practice.

Examples of Regular Astigmatism.—The following will serve as examples of the six varieties of Ast. Their correction will be explained afterwards.

1. (Am.)—Vertical meridian Mc. M. = +2 D. Horizontal meridian Emc. Amount of Ast., or the difference in refraction of the two principal meridians, or Am. = 2 D.

2. (Ah.)—Vertical meridian Emc. Horizontal meridian Hc., H. being −2 D. Amount of Ast. or Ah. = 2 D.

3. (M. + Am.)—Vertical meridian Mc. = +4 D, horizontal meridian Mc. = +2 D. Difference in refraction between the two meridians, or the amount of Ast. or Am. = +4 D −2 D = 2 D.

4. (H. + Ah.)—Vertical meridian Hc. H. = −2·50 D; horizontal meridian Hc. = −7 D. Difference in refraction between the two meridians, or the amount of Ast., or Ah. = − (7 D −2·50 D) = 4·50 D.

5. (Amh.)—Vertical meridian Mc. = +1·50 D, horizontal meridian hypermetropic, H. being = −1·0 D. The amount of Ast., or difference in refractive power of the two meridians, or Amh. =1·50 D *added* to 1·0 D, = 2·50 D.

6. (Ahm.)—Vertical meridian Mc. = +1·0 D; horizontal meridian Hc. = −1·50 D. The difference in refractive power of the two meridians, or the amount of Ast., or Ahm., is, therefore, the same as in example No. 5, viz. 1·50 D +1·0 D = 2·50 D.

In the foregoing examples, the vertical and horizontal meridians have been named as the two principal meridians of refractive defect, because it is in these directions, or nearly in them, that the refractive anomalies in Ast. are most frequently found. But the anomalies of refraction may be found in any other of the ocular meridians, and it is essential to determine the precise inclination of the astigmatic meridians in every instance under notice before the correction of the Ast. becomes possible.

To Determine the Direction of the Principal Ametropic Meridians in Ast. and their Degrees of Ametropia.—Various descriptions of test objects have been arranged for determining the direction of the principal ametropic meridians in cases of Ast. Of these probably the most convenient are, firstly, circles, or semicircles such as the one represented in fig. 54 having radii of lines disposed at certain intervals, and marked by figures at the circumference which indicate the number of degrees into which the circumference is divided by them; or, secondly, series of letters formed by lines placed at various angles of inclination, such as those known as 'Pray's types.'

If a semicircle be employed, a line forming the base line of the semicircular arc is usually marked 90° at each extremity, while the line perpendicular to this base line is marked 0. The intervening radial lines are marked according to the number of degrees of the circumference included between them and the vertical line (see fig. 54). Trial frames usually have the vertical line marked 90° ; the horizontal lines marked 0 on one side, and 180° on the other.

The letters of Pray's types are formed of thick dark lines, with white interspaces of the same dimensions. There are letters with the lines disposed in a vertical direction, others have the lines horizontal, and others with the lines inclined at various angles between these two directions.

In using these astigmatic test objects, the person under observation is placed at a distance of 15 or 20 feet from them, the objects being placed on a level with the face, and so that the light may fall on them. Note is then taken of the particular line, or particular letter, which appears most distinct to the astigmatic eye. If one of the lines or letters is seen without any blurring or confusion, it is apparent that the case is one of the simple form of Ast., and that the principal ametropic meridian is in the direction of the line or letter which is seen clearly. Let the line seen clearly be a vertical line, or the letter of Pray's types which is composed of vertical lines, the vertical meridian of the eye is ametropic, the horizontal emmetropic. If now a + and − cylindrical lens of one dioptric be successively passed before the eye with the axis of the lens held at right angles to the ametropic meridian, it will settle the question as to the nature of the Ast., whether it be Mc. or Hc. In the instance supposed, the axis must be horizontal. If the Ast. be Mc., a −, if Hc., a +, cylindrical lens will render all the other lines of the dial, or letters of Pray's types, more dark and distinct. A succession of lenses of the kind thus indicated may then be passed before the eye until the lens which entirely neutralises the ametropia, shown by all the lines and letters being clearly seen, is met with. The power of this lens being known, the nature and degree of the Ast. will be equally known.

If, on regarding the dial or types, none of the lines or letters appear free from blurring, although some have more definition than others, the case is one either of compound, or mixed Ast. The direction of the line or letters seen least hazily is noted, and the inclination of the meridian of chief refractive defect is then known. Proceeding as before, a weak Sph. + or − lens will indicate the nature of the ametropia in the contrary meridian. The Sph. lens which completely clears the line of indistinctness in the chief ametropic meridian shows the amount of ametropia in

H

the opposite meridian, and at the same time reduces the case to one of simple Ast. A weak + or — Cyl. lens, with the axis perpendicular to the inclination of the chief ametropic meridian, will show the nature of the ametropia in this meridian, and a succession of trials with such lenses will show its amount, by noting the lens which completely clears V. in the opposite direction. If the Cyl. lens correspond in refractive quality with the spherical lens which corrected the ametropia in the chief meridian, the case is one of compound Ast.; if the cylindrical lens be of opposite quality to the spherical lens, the case is one of mixed Ast. Thus supposing, for example, the person under observation sees the horizontal line of the dial or letter formed of horizontal lines least indistinctly, the principal meridian of greatest ametropic defect is shown to be in that horizontal direction, the principal meridian with least ametropia is shown to be vertical. On trying with + and — Sph. lenses of low power, a + lens increases the definition of the horizontal line. The ametropia of the vertical meridian, or meridian of least ametropia is shown to be H. A + 1·25 D Sph. lens completes the definition of the horizontal line. The ametropia of the vertical meridian is shown to be H.=1·25 D. The case is now reduced to one of simple Ast. A low power + Cyl. lens, axis vertical, lessens the indistinctness of the vertical line. The ametropia is therefore shown to be of the same nature in both principal meridians; and the case is shown to be one of compound Hc. Ast. On trial, a Cyl. + 2·50 lens, axis vertical, clears the vertical, and now with it and the + 1·25 D Sph. lens together, the lines in all directions are cleared. The case is shown to be a case of compound Hc. Ast., Ah. vertical = — 1·25 D, Ah. horizontal = — 3·75 D. Or, as another example, the lines in a vertical direction being seen with least indistinctness, and the chief ametropic meridian being shown to be vertical, a + 1 D spherical lens renders the vertical lines quite distinct. The horizontal ametropia is shown to be H.= — 1 D. It is presumed that steps have been taken to prevent, by the use of atropine, the exercise of Acc. on the part of the patient, or otherwise the H. in the horizontal meridian might be neutralised by its exercise. On trial, a — 1 D Cyl. lens, axis horizontal, is found to lessen the blurring of the horizontal lines. The ametropia in the vertical meridian is positive in quality, opposite to that of the horizontal meridian, and the case is shown to be one of mixed Ast. A — 3 D Cyl. lens, axis horizontal, the spherical + 1 D lens being retained, renders horizontal and all the other lines clear. The case is thus proved to be one of mixed Ast., with vertical meridian myopic = + 2 D, and horizontal meridian hypermetropic = — 1 D.

Correction of Regular Astigmatism.—The correction of the first two varieties of regular astigmatism—(1) simple Mc. Ast., and (2) simple Hc. Ast.—when the nature and degree of the ametropia in the ametropic meridian have been determined, is at once effected by a suitable cylindrical lens. The axis of the lens must be placed in the direction of the *visual* defect, or, in other words, of the emmetropic meridian. The examples before given of these two varieties will therefore be corrected as follows :—

Ex. 1.—(Am. = 2 D.) Vertical meridian Mc., + 2 D ; horizontal, Emc. A Cyl. − 2 D lens, with the axis horizontal, corrects the Am., and renders the whole eye emmetropic.

Ex. 2.—(Ah. = 2 D.) Vertical meridian Emc. ; horizontal, Hc. − 2 D. A Cyl. + 2 D lens, with the axis vertical, corrects the Ah., and renders the whole eye emmetropic.

The correction of the compound and mixed forms of Ast. may be effected in either of two ways. The Ast., or the difference between the two principal meridians, may be first rectified, and the remaining ametropia corrected by a suitable Sph. lens ; or the case may be in the first instance reduced to the simple form of Ast., myopic or hypermetropic, as the state may be, by an appropriate Sph. lens, and the Ast. then corrected by a proper Cyl. lens. The latter is probably the shortest method. The degrees of ametropia in the two principal meridians being known, whether in one of the varieties of the compound or of the mixed form of Ast., the correction of either meridian by a proper Sph. lens will leave that meridian emmetropic and the other meridian alone ametropic. The case will then be one of simple Mc., or simple Hc., Ast. If the meridian possessing the relative maximum of refractive power be so corrected, the remaining meridian will be rendered Hc. ; if the meridian having the relative minimum of refractive power be corrected, the remaining principal one will be left Mc. The application of a suitable Cyl. lens will then complete the correction as shown in Exs. 1 and 2.

To illustrate this in the correction of the examples of compound and mixed Ast. already given :—

Ex. 3.—(M. = + 2 D + Am. = 2 D.) Vertical meridian Mc. + 4 D ; horizontal meridian Mc. + 2 D. A spherical − 2 D lens will render the horizontal meridian emmetropic, and reduce the refraction of the vertical meridian to a M. of + 2 D. The case will then be similar to Ex. 1, and a Cyl. − 2 D lens, with the axis horizontal, will complete the correction.

Ex. 4.—(H. = − 2·50 D + Ah. = 4·50 D.) Vertical meridian Hc. − 2·50 D ; horizontal − 7 D. A spherical lens + 2·50 D will render the vertical meridian emmetropic, and leave the horizontal meridian Hc. 4·50 D. The case will then be similar to Ex. 2, simple Hc. Ast., and a Cyl. + 4·50 D lens, with the axis vertical, will complete the correction.

Ex. 5.—(Amh. = 2·50 D.) Vertical meridian Mc. + 1·50 D ; horizontal meridian Hc. − 1·0 D.

A spherical − 1·50 D lens will render the vertical meridian

Emc., and leave the horizontal meridian Hc. − 2·50 D. A Cyl. + 2·50 D lens, with the axis vertical, will complete the correction.

Or a spherical + 1·0 D lens will render the horizontal meridian Emc., and leave the vertical meridian Mc. + 2·50 D. A Cyl. − 2·50 D lens, with the axis horizontal, will complete the correction.

Ex. 6.—(Ahm. = 2·50 D.) Vertical meridian Mc. + 1·0 D; horizontal meridian Hc. − 1·50 D.

A spherical + 1·50 D lens will render the horizontal meridian Emc., and leave the vertical meridian Mc. + 2·50 D. A Cyl. − 2·50 D lens, with the axis horizontal, will complete the correction.

Or a spherical − 1·0 D lens will render the vertical meridian Emc., and leave the horizontal meridian Hc. − 2·50 D. A Cyl. + 2·50 D lens, with the axis vertical, will complete the correction.

The adjoining table shows diagrammatically the six varieties of regular Ast., which have just been described, and will serve to explain further the modes of correction in the several examples of them given in the text.

It is obvious from the foregoing illustrations that two methods can be practised for reducing each variety of compound, as well as of mixed, Ast., to a condition of simple Ast. Either the meridian in which there is the relative maximum, or the one in which there is the relative minimum, of ametropia, may be corrected by a suitable spherical lens in order to effect the reduction. If the minimum be selected, as in the examples given of compound Ast., the remaining ametropic meridian retains its original kind of ametropia, but in a lessened degree. If the maximum of ametropia be corrected, the nature of the ametropia in the remaining principal meridian will be changed, and spherical and cylindrical lenses of opposite refractive conditions will have to be employed in the total correction. In mixed Ast., the remaining ametropic meridian, after the reduction to the simple form, may be either Mc. or Hc., according as one or other of the principal meridians has been corrected to reduce it to the simple form. Generally the mode of correction in which the conditions of refraction are simplest in kind, and least in degree, best answer the object in view. The creation of a new and opposite kind of defect in the remaining meridian should be avoided in all cases when practicable; while the plan which corrects one meridian and at the same time lessens the defect in the other, should be aimed at in all instances in which the reduction of a complicated to a simple form of Ast. is under consideration.

There is a third mode of correcting compound and mixed forms of Ast., by neutralising the ametropia in each of the two principal meridians by suitable Cyl. lenses. The axes of the lenses will have to be placed at right angles to each other. Such bi-cylindrical lenses present practical difficulties in their manufacture and adjustment, and are consequently rarely employed in astigmatic correction.

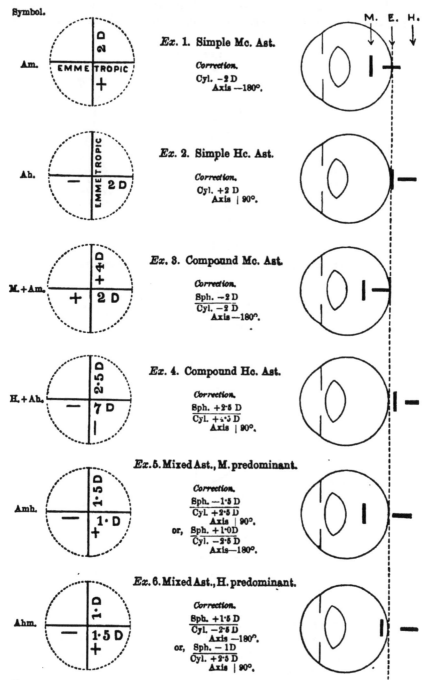

FIG. 56.—DIAGRAMMATIC VIEW OF THE SIX VARIETIES OF REGULAR ASTIGMATISM, WITH THE CORRECTION OF AN EXAMPLE OF EACH VARIETY.

The construction of plano-cylindrical, bi-cylindrical, and spherico-cylindrical, lenses, has been illustrated in a previous

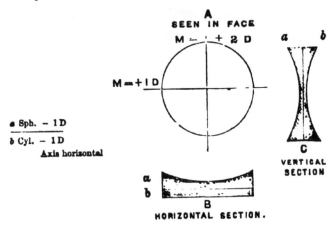

Fig. 57.—Compound Mo. Ast. corrected by a Spherico-Cyl. Lens.

chapter (see figs. 26 to 31). Figs. 57 and 58 will serve to illustrate the application of spherico-cylindrical lenses in the correction of the compound and mixed forms of Ast.

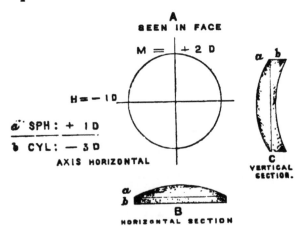

Fig. 58.—Mixed Ast., M. preponderating, corrected by a Spherico-Cyl. Lens.

Diagnosis and Correction of Irregular Astigmatism.—Irregular Ast. being usually due to superficial alterations of form, resulting from inflammatory affections or other lesions of the cornea, some indications of the disease to which the eye has been subjected are the general accompaniments of this form of Ast., such, for example, as cicatrices of corneal ulcers, faults of curvature, corneal opacities, ánd other abnormal conditions, which may be readily observed by lateral illumination of the cornea, or by direct ophthalmoscopic examination.   Not only is the visual acuteness found to be impaired owing to the imperfect transmission of light due to these changes, but objects looked at by the patient are seen deformed.   Straight lines appear wavy, and differ in depth of colour in some of their parts.   Keratoscopy exposes the irregularity of the corneal curvature, by the manner in which the lights and shadows follow each other in succession, when the reflecting mirror is gently rotated on the axis of the handle which supports it.   No forms of lenses, whether Sph. or Cyl., correct the deformity of objects or restore normal acuteness of vision, although the deformity is less, and acuteness of vision frequently improved, when the access of rays of light is limited by the application to the eye of a stenopœic opening.

Influence of Ast. on Military Service.—It may be readily understood from considering the nature and effects of Ast., that it is an optical condition which must seriously interfere with the accurate performance of some of the most important duties of soldiers.   The condition of M. does not prevent the eye from seeing clearly at particular distances; H. interferes with good sight less at some distances than it does at others; but Ast. renders vision more or less indistinct at all distances.   Objects not only appear indistinct, but they are to some extent altered in form; the extent and direction of their disfigurement varying according to the kind and amount of Ast. in the eye which looks at them.   The power of forming correct estimates of the distances of objects is much impaired.   In using a rifle, neither the object aimed at, nor the sights of the rifle itself, can be seen accurately—their outlines are blurred, more in one direction than another.   Moreover, this faulty state of vision is accompanied by an amount of ocular effort and persistent feeling of uncertainty regarding the true characters of objects that add materially to the difficulties of any astigmatic person whose duty it may be to observe objects before him with promptness and precision.   A soldier who is astigmatic cannot, therefore, be relied upon for the proper discharge of his trust if he be placed on guard as a sentry—particularly on active service in the field, where alertness as well as a wide scope of view, and a well-defined recognition of all objects are often of essential importance.   Even in armies in which the use of correcting glasses is permitted, Ast. will always be a disqualifying defect, for it is almost impracticable to make provision for supplying the kinds of lenses, already described, which are necessary for correcting this form of ametropia.

The rules under which astigmatic defects of vision disqualify a recruit from acceptance for military service, and the regulated manner of determining the disqualification, will be found in Chapter IX.

CHAPTER III.

Subjective and Objective Modes of Visual Examination—Objective Assessment of Refraction—DIRECT MODE OF OPHTHALMOSCOPIC DIAGNOSIS—Conditions which Concern the Observer—Those which Concern the Observed Eye—Rays Emitted by an Emmetropic Eye—By a Myopic Eye— By a Hypermetropic Eye —Effects of Foregoing on the Observer's Eye—Application of Observations to Diagnosis—Erect and Inverted Images—Objective Diagnosis of Em.—Of M. —Of H.—Astigmatism—MEASURE OF AMETROPIA BY REFRACTION OPH-THALMOSCOPES—Manner of Employing them—Diagnosis of Degrees of M.— Of H.—On the Use of Refraction Ophthalmoscopes—KERATOSCOPY—Diffi-culties in Military Practice—Appearances in Keratoscopy—Keratoscopy with Concave Mirror—Application in Practice—Diagnosis of Degree of M. or H.— Measure and Correction of Ast.—Keratoscopy with Plane Mirror.

Subjective and Objective Modes of Visual Examination.—The methods of determining the visual acuteness and ocular refractive conditions hitherto described have depended in a great degree on the description by the person under observation of his own visual impressions. The subject under examination has told the surgeon what he could see, or could not see, under given circumstances. Reliance has necessarily had to be placed by the examiner on the honesty and correct description by the person under examination of the subjective phenomena that have occurred to him. If the person who has been tested has only been actuated by a desire to assist the examiner in his investigations, and has answered the tests put to him intelligently, the conclusions arrived at will be correct and sure ; but if, intentionally or otherwise, the replies given to the examiner and the statements made by the person examined have not been in accordance with the person's real impressions, then the conclusions arrived at may be false. The accuracy of the conclusions may be tested by increasing the number of tests and varying their nature ; but these proceedings occupy more time than frequently can be spared for the purpose. If, therefore, modes of examination can be practised in which there is no need for the participation of the person under examination, in which the surgeon's own observations of the eye and its appearances under certain conditions will suffice to reveal to him its refractive quality, and, if ametropic, its amount of ametropia, it is evident that such sources of fallacy as have just been referred to will be altogether removed. If, moreover, satis-factory results can be attained rapidly by an objective examination of the kind alluded to, then obviously the gain will be very great, both to surgeons and patients, and especially to surgeons who have to deal with children who cannot readily describe their sensations, or who have to attend to large numbers of patients within limited periods of time, as often happens in civil hospital practice. It is not to be wondered at, then, that during the last few years a great

amount of attention has been given to objective methods of ex-
ploring and determining ocular states of refraction, not only generally in civil life, but also in military practice in countries where
the system of conscription is in force, and in which the numbers
of men to be examined, and the amount of false assumption or
exaggeration of visual defects are far greater than British military
surgeons ever have to deal with. As a general rule, a large proportion of conscripts for Continental armies would avoid military
service if they could, and hence naturally have a tendency to ex-
aggerate defects when they exist, or to make it believed that they
exist when they do not exist, while, as a general rule, in a system
of voluntary enlistment, just the reverse holds good, and the recruits
do their best to show their fitness for service. It is obviously comparatively easy for a surgeon to determine whether a recruit's power
of sight is up to a given standard when the man is doing the
utmost he can to show his sight at the best; while it is a very difficult matter to prove that the quality of vision of a conscript is up
to the required standard, when the man is trying to counteract the
surgeon's efforts, unless the surgeon is sufficiently expert to determine the question by objective observation.

Objective Assessment of Ocular Refraction.—Ophthalmoscopic
observation of the fundus affords very useful assistance to surgeons
in enabling them to establish a diagnosis of the refractive condition
of an eye in doubtful cases. If the subject's vision be really defective owing to ametropia, the presence and kind of ametropia, after
a little practice, and approximately the degree of the defect, can
be ascertained by its means. The determination of the degree of
V. by methods previously described will indicate the amount of the
disability stated to exist, and as, if any attempt at simulation be
practised, a high degree of defect is likely to be assumed, the ophthalmoscopic examination will resolve all doubt on the subject,
without the person under observation having any power to interfere
with the conclusion at which the surgeon may arrive.

Both the direct and indirect methods of ophthalmoscopic observation can be used for ascertaining objectively the refractive state
of an eye, but the former method is more reliable and useful. The
direct mode of examination in which the mirror alone is employed
can be applied in two ways, viz. by using it for observation of an
illuminated part of the fundus, or particular object in this illuminated portion; or by using it for observation of the shadow by
which the illuminated part is bounded. This latter application of
direct examination is known as ' *Keratoscopy*,' or sometimes ' *Retinoscopy*.'

In order to judge not merely the quality of refraction, but to
determine the degree of ametropia when ametropia exists, other
expedients, in addition to the simple use of the concave mirror, are
necessary. The expedients in ordinary use are mechanical arrange-

ments for applying any required correcting lens to the aperture of
the mirror by means of '*refraction ophthalmoscopes,*' or the application, as in keratoscopy, of a series of trial lenses in front of the eye
under examination.   An explanation of each method follows.

Assessment of Ametropia by Direct Ophthalmoscopic Observation of the Fundus.

Ophthalmoscopic Diagnosis—Direct Method.—In this proceeding
the ordinary concave mirror alone is employed.   It is understood
that the examining surgeon is himself emmetropic, or renders his
eye emmetropic by correction through a suitable lens placed in the
clip at the back of the mirror.   The surgeon arranges himself so
that the mirror, through the sight-hole of which he makes his observations, is at a distance of 18 inches or more from the eye under
observation, the light having its usual relative position to the patient
and observer.   For precise observation, especially in cases where
there is reason for suspecting an attempt at simulation, it is well
to place the patient under the influence of atropine.   If this should
not be convenient, or not considered desirable, then the eye of the
patient should be directed to some distant point so that its accommodation may be relaxed as far as possible during the examination.

Rays Emitted by an Emc. Eye.—In applying this method it is
to be remembered that the rays reflected from the illuminated

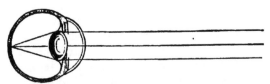

FIG. 59.—RAYS EMITTED BY AN EMMETROPIC EYE, ACCOMMODATION BEING AT REST.

retina of an emmetropic eye whose Acc. is at rest have such directions given to them in traversing the successive refractive media
that they pass outwards from the eye as parallel rays—that is, all
the reflected rays emitted from each point of the retina after leaving the eye travel in a direction parallel with the axial ray belonging to that point (see fig. 59).

Rays Emitted by a Mc. Eye.—The rays reflected from the retina
of the *myopic* eye, when its Acc. is at rest, in issuing from the eye
follow the same paths as the diverging rays which enter it from its
distant point of distinct V., and therefore pass away from the
globe as convergent rays, that is, all the rays from any one given
illuminated point of the retina, instead of being parallel with the
axial ray belonging to that point, as happens with the Emc. eye,
converge towards it and meet upon it at some point.   This meet-

ing point is the remote point of distinct vision of the eye under observation, and its distance from the eye will depend upon the degree of M. (see fig. 60). The higher the degree of M. the more the rays converge, and, as a result, the nearer to the eye their meeting-point. The point on the retina from which the reflected rays start, and the point at which they meet in front of the eye are conjugate foci and in the case of a retinal object being regarded, a real and inverted image of that object will be formed in the position of the conjugate focus situated in front of the eye. This aërial image will be larger than the object from which it is projected.

FIG. 60.—RAYS EMITTED BY A MYOPIC EYE, ACCOMMODATION BEING AT REST.

Rays Emitted by a Ho. Eye.—The rays reflected from the retina of the Hc. eye pass out of the globe as divergent rays, that is, instead of approaching toward the axial ray as described in the Mc. eye, they diverge from it. The greater the deficiency of refractive power in the eye under observation, or, in other words, the higher the degree of H., the greater will be the angle of divergency from the axis of the cone of rays (see fig. 61). If these emitted rays were prolonged backwards with the angle of divergency unchanged, they would meet at a point behind the retina, and, at the distance of this point, an erect, enlarged, virtual image of an illuminated retinal object would be formed.

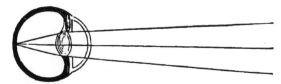

FIG. 61.—RAYS EMITTED BY A HYPERMETROPIC EYE, ACCOMMODATION BEING AT REST.

So far then as an eye looking through the sighthole of the mirror is concerned, the rays falling on it will be parallel if the eye under observation from which they proceed be Emc., convergent if they come from a Mc. eye, and have not yet reached their meeting point; and divergent if they come from a Mc. eye after they have met and crossed, or if they come from a Hc. eye.

Effect of Rays from an Emc. Eye on the Observer.—The observer being understood to be Emc., the effect produced on his vision by rays having the several directions named, will vary according as he exerts Acc. or does not exert it.   Being emmetropic, if no Acc. be exerted, his eye is adapted for focussing parallel rays and he will be able to see objects on the retina of the Emc. eye under observation ; but if Acc. be exerted and his V. adapted to the distance of the retinal object, he will not be able to see it with definition, because his eye will then be adapted for focussing divergent rays, while the rays emitted by the Emc. eye and falling on his own are parallel rays (see fig. 62).

FIG. 62.—RAYS REACHING THE OBSERVER'S EYE FROM AN EYE WHICH IS EMC.

Effect of Rays from a Mc. Eye on the Observer.—If the eye under observation be Mc., the clearness of view of the fundus will not only depend on the fact of Acc. being exerted or otherwise, but will also be influenced by the relative position of the examining surgeon and the position of the punctum remotum of the eye under examination.   If the observer be placed so as to be within the remotum of the Mc. eye under examination, and the emitted converging rays have not met and crossed before reaching the observer's eye, a clear image of a retinal object cannot be formed on the observer's retina.   His eye is adapted for focussing parallel rays if no Acc. be exerted, and will be adapted for divergent rays, if Acc. be exerted, while the rays reaching it under the conditions named are convergent rays.   If the observer be placed so as to be outside and beyond the remotum of the observed eye, so that the emitted converging rays will have met and crossed before arriving at the observer's eye, then, under such circumstances, by accommo-

FIG. 63.—RAYS REACHING THE OBSERVER'S EYE FROM AN EYE WHICH IS MC., THE RAYS HAVING CROSSED.

dating for the distance of the meeting point of the rays from the observed Mc. eye, he will see the image formed at this point as distinctly as if it were the object itself instead of its image.   The emitted rays having crossed, fall on the observer's eye as diverging rays, and by the assistance of a suitable amount of Acc. they are brought to an exact focus upon his retina.   The higher the degree of M. the sooner will the emitted rays meet, and the nearer to the eye will be the aërial image of the fundus (see fig. 63).

Effect of Rays from a Hc. Eye on the Observer.—If the observed eye be Hc., as already mentioned, the emitted rays have a divergent direction, and fall upon the observer's eye as divergent rays. If the observer accommodates for the apparent origin of these divergent rays, that is for the virtual remote point of the eye under observation, he will be able to bring the divergent rays to a focus on his retina and thus obtain a clear view of the objects on the fundus of the observed eye. The divergence is neutralised by the observer bringing his Acc. into action. If the observer cannot exercise sufficient Acc. for the purpose, or, in other words, if the virtual remotum of the observed eye be within the near point of distinct vision of the observer, clear vision of retinal objects on the fundus of the Hc. eye will not be obtained (see fig. 64).

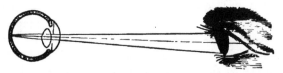

Fig. 64.—Rays reaching the Observer's Eye from an Eye which is Hc.

Application of Observations to Diagnosis.—The application of the observations just described suffice to show whether the eye under examination is emmetropic or ametropic. The fact that no object on the fundus is defined, if any Acc. be exerted, proves that the eye is emmetropic ; the fact that objects are defined when Acc. is exerted, proves that it is ametropic, and further, if the eye be Mc., that the remotum of the Mc. eye is within the range between it and the distance of the observer. But more precise knowledge of the nature of the image seen by the observer is necessary in order to define the exact nature of the ametropia. This information is obtained in the following manner.

Although the observer takes a position at a distance of a foot and a half to two feet from the patient at starting, he may find that he has to vary the distance considerably before obtaining a clear view of the fundus. He will then move himself backwards and forwards, so as to increase or lessen the distance, according to need, until a clear view of a part of the fundus is obtained. If the eye under observation be Emc., and the observer finds a difficulty in relaxing his Acc., he will have to bring the mirror close to the eye of the patient before he can obtain a view of the fundus. When the observer's eye is brought thus close, all Acc. relaxes spontaneously, and the parallel rays emitted from the Emc. eye can then be focussed on the retina of the observer as easily as if they proceeded from an object at infinite distance.

Erect and Inverted Images.—The observer, as soon as he has obtained a clear view of the portion of the fundus before him, fixes upon a retinal vessel as a mark of observation. A vessel proceeding from the optic papilla forms a convenient mark, as it is conspicuous by contrast with the background over which it passes. Moving his head to one side, the observer watches the apparent movement of the retinal vessel he has selected as a mark. If the object regarded follow in its apparent movement the same direction as the head of the observer—if the observer on moving his head to the right sees the vessel move to his right also, he knows that he is looking at an upright image; if, on the other hand, the object in its apparent movement follows a direction opposite to that of the observer's head—if the observer on moving his head to the right sees the vessel move to the left—he knows that he is looking at an inverted image of the object he has taken as a mark.

Objective Diagnosis of Em.—Applying the observations just described, if the observer, accommodating for the distance of the eye under examination, fail to obtain a distinct image of a retinal vessel, the eye may be regarded as neither Mc. nor Hc., but Emc. The conclusion will be rendered complete if the observer, carrying the mirror close to the eye he is examining, and no longer exerting Acc., is then able to see a clear enlarged image of a portion of the fundus, and if this image proves itself to be an erect image by moving in the same direction as the observer moves his head.

Objective Diagnosis of M.—If the observer, with the mirror at 18 inches or farther, sees a distinct image of a portion of the fundus, and, marking a particular vessel, finds it move in the opposite direction to that in which he moves his head, it is evident that he has a Mc. eye before him. The image that he is seeing is an inverted image formed at the remotum of the eye under examination. If the observer's near point of distinct vision is beyond the range of distance between his eye and the image, the image will not be so distinct as it will become when he moves himself farther away from it. If the remotum of the Mc. eye be farther off than the distance at which the observer can usefully employ the mirror for illumining the eye—a distance of 3 feet or more—no distinct image will be seen, and if a vessel can be distinguished it will move in the same direction as the observer's head, for the reflected rays emitted from this feebly Mc. eye will not have crossed before they have reached the eye of the observer.

Objective Diagnosis of H.—If the observer with the mirror at 18 inches or farther sees an image of a portion of the fundus while accommodating for its apparent distance, and, marking a particular vessel, finds it move in the same direction as that in which he moves his head, the conclusion is that he has a Hc. eye before him. The image is seen more distinctly as the observer approaches the eye under examination until he has arrived at 5 or 6 inches from it. The objects seen do not appear so enlarged as they do in the Emc. eye when the observer's eye is brought close to the observed eye, but the chief distinction between the condition of Em. and H. is to be found in the fact that no image can be seen in the Emc. eye at a distance of 6 or 8 inches, or further off, unless all accommodation is suspended, while in the Hc. eye the image is seen when Acc. is exerted in proportion to its distance and the degree of the defect.

Objective Diagnosis of Ast.—The same indications may be turned to account in determining the existence of Ast. The indications will vary in the two principal meridians in kind and degree, according to the nature and amount of Ast. presented. In mixed Ast. the indications will be contrary in one meridian to what they are in the opposite meridian; they will correspond with those which characterise the Mc. condition in one meridian, while they will agree with those distinguishing the Hc. condition in the opposite meridian. If no Ast. be present in the observed eye the retinal vessels radiating over the fundus are seen in all directions with equal distinctness.

ASSESSMENT OF AMETROPIA BY REFRACTION OPHTHALMOSCOPES.

Description of Refraction Ophthalmoscopes.—These ophthalmoscopes are used for determining objectively the state of refraction of an eye on the same principles as those on which the direct method of observation just described by the concave mirror alone is based; but the mirror is furnished with a series of small convex and concave lenses with a view to determine not only the nature of any ametropia which may exist, but also to estimate its amount by ascertaining the power of the lens necessary for its correction. There are many forms of refraction ophthalmoscopes, differing in mechanical details, but in all the varieties the small correcting lenses are so arranged that they can be moved in rotation behind the mirror in such a way as successively to cover the sighthole through which the observer looks at the eye under examination. There are usually two superimposed discs behind the mirror, each carrying a certain number of lenses, arranged circularly near its outer border, and each capable of being rotated independently into any position at the discretion of the operator. I have found the refraction ophthalmoscope arranged by Dr. G. L. Johnson, of the Royal Westminster Ophthalmic Hospital, easy of manipulation. It is sold as 'Johnson's improved ophthalmoscope' and can be employed satisfactorily either for the direct and indirect modes of ophthalmoscopic examination or for keratoscopy. The lenses are so arranged that in applying them only one refracting medium intervenes between the eye of the observer and the patient in all the lenses between -1 D and -10 D inclusive, and between $+1$ D and $+8$ D inclusive, while by the intervention of a second lens all the intermediate half-dioptrics and higher powers up to -29 D on the one hand, and $+23$ D on the other hand, can be obtained. By this means, although twenty-one lenses are inserted in the two discs, a range of seventy lenses is obtained. Means are adopted for indicating at the back of the instrument the power of each lens brought before the aperture of the mirror. This ophthalmoscope, like others of the same class, is also supplied with two concave mirrors, one small and of short focus, the other large and of long focus. The small mirror has a diameter of seven-eighths of an inch and a focal length of 3 inches, the larger mirror a diameter of 1·4 inch and a focal length of 18 inches. They are so connected and pivoted that either can be instantly brought into position as required. One forms a convenient mirror for direct examination close to the patient's eye; the other is suitable for concentrating light on the fundus when used at a distance from the patient, as in the indirect method of examination or in keratoscopy. The reflecting surface of the smaller mirror is not level, having its plane parallel with the plane of the revolving discs, like the larger mirror, but is set at an angle of 35°, and being made to rotate round its own axis can be turned to reflect the light from and in any direction required. The examining surgeon, instead of inclining the mirror, in order to obtain the fullest amount of light for reflection upon the eye of the patient, can effect the same purpose by rotating the mirror in its position, while he avoids the inclination which would otherwise be given to the correcting lens relatively to his own line of sight. Whatever the position of the inclined mirror, the principal axis of the lens used for correction, and the visual axis of the observer, remain in one and the same line. The illustrations on page 113 show the arrangements of the parts which have just been described.

Manner of using Refractive Ophthalmoscopes for Direct Observation.—The small mirror is brought uppermost. As it is inconvenient for the observer to use one of the disc lenses attached to the mirror for correction of his own ametropia in case he is ametropic, he must either wear suitable spectacles or take his own error of refraction into account when he finds the lens which appears to correct the ametropia of the observed eye. If the observer be Mc. he must subtract the power of the lens equivalent to his degree of M. from the lens which corrects an observed Mc. eye or must add it to the power of the lens which corrects an observed Hc. eye. If the observer be Hc., he will have to deduct the power of the lens representing his degree of H. from the power of the correcting lens of a Hc. patient; he will have to add it to the power of the correcting lens of a Mc. patient. It is necessary also that the Acc. of the observer as well as of the eye under observation should be in abeyance. Holding the instrument upright, he then places the mirror as near to the eye to be examined as can be managed, and rotates the mirror until he has so adjusted it as to concentrate the light on the fundus of the eye under examination. At this close distance to the examined eye there will be little, if any, difficulty as regards the exercise of Acc.

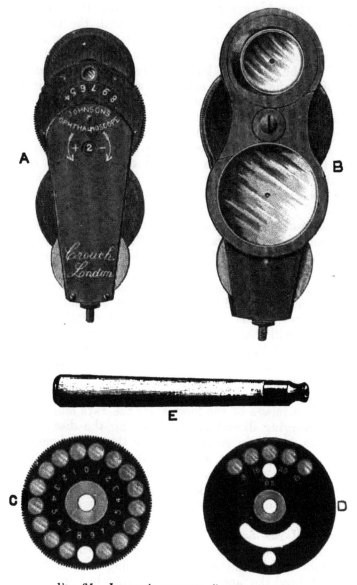

FIG. 65.—JOHNSON'S IMPROVED OPHTHALMOSCOPE.

A, back of instrument, showing portions of the discs carrying the lenses ; B, front, with the two mirrors ; C and D, discs, with lenses ; E, hand e.

by the examiner, and his eye will be adapted for focussing parallel rays. The light should be placed to the outer side of the eye to be examined, and about 6 inches behind it. For this observation it is less embarrassing for the examiner to place himself on the same side of the patient as the eye to be examined—on the right side for the right eye, on the left side for the left eye. The patient should be caused to turn his eye upwards and inwards in such a direction as will bring the optic entrance in line with the observer's line of sight, and to look as far away as practicable in order to keep his Acc. relaxed. The handle of the instrument should be held in the hand of the observer, but his forefinger should rest on the margin of the wheel by the action of which the principal disc is turned, when, if it be the left hand, by pressing the disc from above downwards, or from left to right, the correcting lenses will be brought into position in an ascending series, and by pressing it in a reverse direction in a descending series. The value of each lens as it comes into place is shown on the back of the instrument ; the figures being coloured red for the convex, white for the concave lenses (see fig. 65). The indications will then be as follows :—

Diagnosis of Em.—An enlarged and upright image of the optic entrance and retinal vessels is seen. The details of the image are clear, and are not rendered more clear by the intervention of any + or − lens. It is necessary to be sure that the eye is not Hc., for in the Hc. eye the image may also be clearly seen by the observer if he use his Acc. and so focuses the emitted divergent rays. If, however, the eye is Emc. the intervention of a + lens of low power will change the direction of the parallel rays as they fall on the eye of the observer into converging rays, and the Emc. observer will then fail to get an image as clear as he did when the rays fell on his eye as parallel rays.

Diagnosis of Degree of M.—The observer sees only a blurred and indistinct image of the fundus, for the rays falling on his eye have a converging direction. On rotating the disc and bringing the concave lenses in succession over the sight-hole, the image becomes gradually clearer, and is at last rendered distinct. The refractive excess of the Mc. eye is then neutralised, and the rays emitted from it after passing through the correcting lens fall on the observer's eye as parallel rays. They are changed into the same direction as if they were received directly from an Emc. eye. The power of the first, or least dispersing, lens which produces this result expresses inversely the degree of the M. The image may remain distinct, although a dispersing lens of higher power is placed behind the aperture of the mirror, but in this case part of the action of the higher concave lens is neutralised by the observer exercising a proportionate amount of his Acc.

Diagnosis of Degree of H.—The image of the fundus is seen clearly with the mirror alone, if the observer can exercise Acc. so as to adapt his sight to the apparent distance from which the divergent rays emitted by the eye under observation take their start. The observer tries the intervention of a + lens of low power. If the image remains clear the eye is Hc.; the divergency of the rays emitted by it is lessened, but the observer still accommodates for the apparent distance of the virtual remotum of the eye under observation, and so obtains a clear image of the fundus. The lenses are successively increased in strength until the image of the fundus is rendered more or less indistinct. The rays falling on the eye of the observer are now not divergent nor parallel, but have been rendered convergent. The refractive deficiency of the eye under observation has been more than supplied, and the eye rendered slightly Mc. The lens expressive of the degree of H. will therefore be either the last used, the strongest converging lens under which the fundus was seen clearly, or will be between it and the lens which has rendered the image of the fundus more or less confused.

Remarks on the Use of these Instruments.—Much practice is necessary before an estimate of the degree of ametropia can be formed even with approximate exactness in some instances. Errors are apt to arise from the observer not relaxing his Acc. completely or from failing to determine the precise limits of the maximum of distinctness of the images seen by him—at what point the distinctness is greatest, or at what other point the distinctness begins to be lessened. Other sources of error occur in this method of examination. Theoretically the correcting lens ought to be situated at the anterior focus of the eye under observation, about half an inch in front of the cornea, but from the form of the instrument and the position in which it must be held for reflecting the light upon the pupil, the lens behind the mirror is practically an inch or so beyond this distance. The greater the distance of the lens from the anterior focus of the eye, and the higher the degree of ametropia of the patient, the less precise will be the estimate furnished by the power of the correcting lens.

The allowance to be made on account of the various sources of error just referred to can only be arrived at after long practice with the instrument. The difficulties are increased when the refracting ophthalmoscope is employed for determining the relative differences of refraction in a case of Ast.

ASSESSMENT OF AMETROPIA BY KERATOSCOPY.

Keratoscopy.—Keratoscopy is a name given to a mode of determining the refractive condition of an eye, which is now much resorted to, as it can be employed independently of active assistance on the part of the patient, can be accomplished rapidly, and the end attained in most cases with adequate precision, notwithstanding the presence of amblyopia. Keratoscopy does not depend upon observation of the cornea itself, as the term would seem to imply, but upon variations in the aspects and relative movements of certain lights and shadows seen within the pupillary disc under special ophthalmoscopic arrangements. The term Retinoscopy is sometimes applied to the same method of observation, but this term fails to distinguish it from other modes of ophthalmoscopic observation of the retina, and the meaning of *keratoscopy*, though faulty as an explanatory expression, is universally understood.

Keratoscopy in Military Practice.—The practice of keratoscopy requires not only a mirror, either concave or plain, but also a complete case of trial lenses. As such trial cases do not form part of the regular equipment of military hospitals, military surgeons will rarely be able to avail themselves of this method of observation for deciding the qualities of vision in the cases submitted to their

FIG. 66.—ARRANGEMENTS NECESSARY FOR KERATOSCOPY.

The simple lines show the incident, and the dotted the reflected rays. These are, of course, not seen in nature.

judgment, but must resort to other proceedings described elsewhere in this manual. Whenever trial cases of lenses are available, however, by private means or otherwise, keratoscopy, when the art has once been acquired, will be found to possess many advantages, especially in cases complicated with amblyopia or astigmatism. To be expert in the application of keratoscopy the surgeon must be prepared, as in other modes of estimating amounts of error of refraction objectively, to devote considerable time and attention to its practical acquirement.

Arrangements necessary for Keratoscopy.—The pupil of the eye to be examined having been dilated, and the Acc. paralysed, by the use of atropine, when its employment is not for any special reasons objectionable, the patient, seated in a dark room with the lamp placed well behind him, is directed to look at a distant object just above the surgeon's head. The surgeon, having previously corrected any error of refraction peculiar to himself by a suitable lens at the sight-hole of the mirror, seats himself at a distance of about 3 or 4 feet from the patient, and lights up the eye to be observed by means of the mirror held vertically. The most convenient mirror for the purpose is a concave mirror whose principal focus is at a distance of from 1 foot to 18 inches. The illustration hardly indicates distance enough between the surgeon and the eye under observation.

Appearances in Keratoscopy.—Observing steadily the illumined intra-pupillary disc, the surgeon rotates slowly the mirror on the axis of its handle, and on doing this the appearance of a dark shadow may be noticed at one of the lateral borders of the red disc, and, if the rotation be continued, the shadow will advance until it obscures the whole space of the disc. At a certain interval half the disc will appear illumined and half appear shaded (see fig. 67). If it be the half of the disc to the left of the observer which is obscured, while the half to his right is illumined, the shadow is

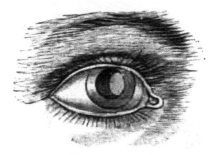

FIG. 67.—APPEARANCE N KERATOSCOPY. ADVANCE OF THE SHADOW.

advancing from left to right, as seen by the observer; if the shadowed and illumined portions be reversed in position, the shadow travels from the observer's right towards his left hand, so that the shaded part always marks the side from which the shadow travels, and the illumined part that towards which it is travelling. The exposed border of the shadow will sometimes be straight, sometimes more or less crescentic in outline; its intensity also, and rate of movement with reference to the movement of the mirror, will vary in different cases.

It will be seen from the arrangement of the light and from the distance at which the observer holds the concave mirror, that the converging rays reflected from the mirror will meet and intersect in the air, and that, having crossed, they will then fall on the eye as diverging rays proceeding from the point of intersection, and will ultimately be brought to a focus, or will approximate to a focus, and will form a more or less defined image on the retina according to the refractive quality of the media of the eye under examination by which they have been acted upon in their passage. The part of the retina outside the illuminated image will be in shadow. It is this shadow, dark by contrast with the illuminated part of the fundus, the movements of which, after the image and its surrounding shadow have been acted upon in returning through the refractive media of the eye towards the eye of the observer, the observer examines in keratoscopy (see fig. 68). The more true and defined the image on the retina, the brighter will be its edge and the darker and more distinct the border of the shadow ; on the other hand, the more the image is confused by circles of diffusion, the more weak will be its illumination though more extended in area, and the less defined and the more crescentic in outline will be the border of the shadow. It follows, therefore, that it is in

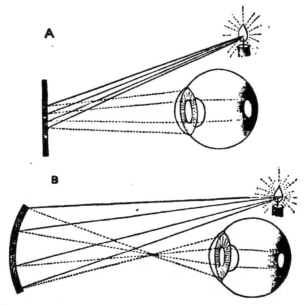

FIG. 68.—NATURE OF THE LIGHT AND SHADE OBSERVED IN KERATOSCOPY.
IRIDIAL SHADOWS.

A, plane mirror ; B, concave mirror beyond focal distance. Incident rays diverging in both instances, but only crossed in the latter.

the lower degrees of ametropia, where there is least diffusion, that the most defined shadows are presented to view, and, *vice versâ*, that in the higher degrees of ametropia the shadows are less marked in character and outline.

Nature of Keratoscopy.—It will be observed from what has been stated that the direction in which the light and shade alternate, or, in other words, the direction from which the shadow comes and to which it goes, does not depend solely upon the direction of movement of the mirror, but, with the same movement of the mirror, will vary according to the refractive quality of the eye under observation (see fig. 69, c, 1, and c, 2). In keratoscopic observation, therefore, the surgeon first takes into account the direction in which he moves his mirror, and secondly, the direction in which the shadow follows the reflected light within the pupillary disc ; whether the path of the shadow is direct, or follows the light in accordance with the movement impressed on the mirror, or whether its path is inverted, the shadow following the pupillary

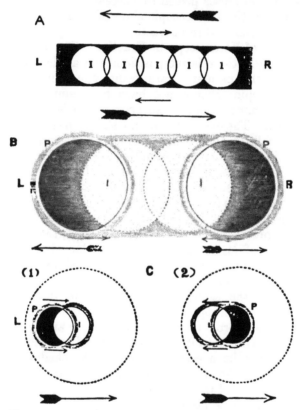

Fig. 69.—Explanation of Appearances in Keratoscopy and Application to Practice.

A, lateral movement of the shadow of the iris with pupillary aperture, in opposite direction to that of the mirror ; B, course of the shadow in the pupillary disc (P), contrary to the movement of the light ; C, effect of difference in the refracting media of the eye : (1) movement direct, image on retina inverted, M. higher than 1 D ; (2) movement contrary, Em. or H. ; image on retina direct, H., Em. or M. of low degree. The feathered arrows show the movement of the light, the plain arrows the course of the shadow ; L, left ; R, right ; P, pupillary disc ; I, retinal image.

light in the opposite direction to that of the movement given to the mirror.

Estimating Ametropia by Keratoscopy.—These constitute the principal points to be noted so far as concerns the determination of the refractive quality of the eye under observation ; but, if the eye be ametropic, certain other points already alluded to have to be noted in forming an estimate of the amount of the ametropia. The form of the edge of the shadow, whether linear or crescentic; the kind of movement, whether slow or relatively quick; the degree of brightness of the illuminated portion of the disc, and the definition of the edge of the shadow; all assist in affording means of estimating the amount of ametropia. There is, however, only one method of determining the degree of ametropia with precision, viz. by the use of trial lenses placed in succession before the eye under observation until the keratoscopic signs of the ametropia are caused to disappear or changed in character.

Application of Keratoscopy in Practice.—The surgeon, patient, and light being in the relative positions and at the distances before described, and the patient's gaze being fixed on a distant object, especially if atropine should not have been employed to prevent exercise of accommodation, the surgeon proceeds to light up the eye to be observed. The eye of the patient should be so turned that the light neither falls on the macula lutea nor on the optic papilla, but to the inner side of both or on the retina below. The mirror is now rotated, and the order of movement of the light and shadow in the pupillary disc watched. The following will be the indications according to the nature of the movements observed :—

1. If the light and shadow follow the direction of the mirror, or in other words if their movement be direct, the eye is myopic, and the degree of M. indicated is higher than 1 D. The mirror, held at a distance of 3 or 4 feet, is outside and beyond the distant point of distinct vision of the examined eye.

2. If, under the same circumstances, the light and shadow move in a direction opposite to that of the movement of the mirror—in other words, if their movement, is an inverse one—the eye is probably Hc., but either Em., H., or M. of a low degree may be present. If it be a case of weak M. the shadow is light, not strongly marked; the mirror is within the distance of the distant point of distinct V. of the eye under observation; the degree of M. is lower than 1 D (see fig. 69).

To Ascertain Amount of M. or H.—A trial spectacle frame is placed on the patient, and, according to the nature of the ametropia diagnosed, either + or − lenses are placed in succession before the patient's eyes, until the shadow is caused to reverse its direction under the keratoscopic observation or becomes hardly noticeable. If the shadow has its direction reversed, the ametropia is more than neutralised, and weaker lenses must be applied until this reverse movement ceases, and the shadow is simply deprived of definition. The lens which produces this result gives approximately the measure of the ametropia. If the result is not similar in opposite meridians of the eye, astigmatism exists, and each meridian will have to be dealt with separately in the manner just described.

Estimating Degree of M.—It will be observed that in diagnos-

ing and estimating M., the distance of the observer relatively to
the distant point of distinct vision of the eye under observation, or
the degree of M., has an important bearing in the keratoscopic
appearances.   If the observer holds the mirror farther off than the
remote point of V. of the eye under observation, the movement of
the shadow is direct, as already mentioned; but if the observer
were to place himself within the distance of the remotum of the
eye under observation, the movement of the shadow would be reversed; while if he were to place himself at the precise distance
of the remotum, no defined shadow would be observable.

If, therefore, the observer place himself at 1 metre distance or
beyond, and the shadow move directly with the mirror, the eye
under observation has M. higher than 1 D, while if the shadow
move inversely as the mirror moves and the eye is Mc., the degree
of M. must be less than 1 D.   In the higher degrees of M. the
refractive condition is easily recognised; in the lower degrees of
M., such as M. of 1 D or lower, the M. is not readily distinguished
from Em. or a low degree of H.; for the movements of the shadow
bounding the image formed at such a distance from the eye under
observation are hardly recognisable.   But if on placing before the
observed eye a + lens of low power, such as one of half a dioptric,
the direction of movement of the shadow is seen to be changed—
from moving inversely to moving directly—there is no doubt the
case is one of low M. and that the small addition of the + 0·5 lens
has given it the usual character of M. of a higher degree.   If it
were a case of H., so small an addition would not have sufficed to
neutralise the Hc. deficiency of refraction.

Estimate and Correction of Ast. by Keratoscopy.—The correction
of Ast. by keratoscopy is effected on the same principles in each of
the meridians, but the process is one which requires considerable
practice in order to accomplish the desired results with accuracy.
If a shadow in a vertical direction differ in definition or intensity,
or show a different rate of movement from the shadow in the
horizontal direction, or if the shadows assume an oblique direction,
it may at once be assumed that Ast. exists which requires correction.   Each meridian is corrected separately, and the eye afterwards tested with the combined spherical and cylindrical lenses
before it, and when the ametropia appears to be neutralised in all
directions, it may be assumed that the glasses are correct.   If the
shadows are not vertical and horizontal, but have an oblique direction when the mirror is moved in the ordinary way, the obliquity
of the astigmatic meridians will be in a corresponding direction.

Keratoscopy with the Plane Mirror.—If under similar circumstances to those just described the surgeon uses a plane mirror at a
distance of 3 feet or beyond, on observation of the eye under examination the effects will be found to be the reverse of those met with
when the concave mirror was used.   The shadow will follow the

same path as if the observer were watching the retina itself, or vessels on the retina, and observing the manner in which these objects move relatively to the movements of his own head in direct illumination of the observed eye. If the eye under examination be Mc. and the M exceed 1 D, the shadow will move inversely as the mirror is moved; if the shadow move directly in accordance with the movements of the mirror, the eye is either Emc., Hc., or is Mc. in inferior degree to 1 D. In the Emc. condition the shadow has only moderate saturation, though it may be well defined; in H. the saturation and definition of the shadow are increased in proportion to the increase in degree of H.; in M. the definition and depth of the shadow also vary according to its degree. With M. higher than 1 D, the image of the illumined portion of the fundus and the shadow, which is a real image formed between the mirror and the observed eye, is well marked; with M. lower than 1 D, the image is erect and virtual, not formed in front of the mirror, and the shadow is wanting in distinctness and saturation.

CHAPTER IV.

Accommodatory Function of the Eye.

The conditions previously described, viz. those of Em., M., H., and Ast., depend upon persistent qualities of eyes. It may be supposed that the quality of an eye, so far as concerns either one of the conditions just named, would remain the same whether the eye, if physically unchanged, be a living eye, or whether it be disconnected from the human body. It is simply a quality which is determined by the forms and refractive powers of the several dioptric media of the eye acting in concert, or of the whole of them together regarded as a single lens, relatively to the position of the retina.

The eye has the power, however, under certain circumstances of changing its refractive adjustment. If it were not so, the eye would be adapted for clear vision at one distance only, viz. that of its distant point. The Emc. eye would only see objects clearly at infinite distance; the Mc. eye at its limited distant point; the Hc. eye would not see objects clearly at any distance. The function, by means of which the eye is able to see objects clearly at different distances, is called the *accommodatory function*, or, briefly, *accommodation*. The discharge of the function constitutes what is known as the *act of accommodation*.

Act of Accommodation.—This act is performed on all occasions when vision is transferred from a relatively distant to a nearer object. If two pages of print are placed a few inches apart before the eye, with an opening in the nearer page, the eye while reading through this opening the print on the more remote page is unable to recognise any of the words on the near page. To read the words on the near page, although they are closer, the eye, in common language, must 'look at them'; and on reading them the eye ceases to be able to recognise the words on the more remote page, athough they are within reading distance and may be directly opposite to the eye. The same effects may be noticed in looking at a landscape through a pane of glass, and then at some small object on the glass itself. The term 'looking at the object' here signifies performing the *act of accommodation*. If carefully observed,

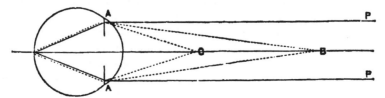

Fig. 70.—Demand for Acc. (see explanation in text).

the act of accommodation will be noticed to be accompanied with a sense of muscular effort. If the parts concerned in the proceeding happen to be inflamed, as occurs in iritis and other morbid states, the act of accommodation is accompanied with considerable, often acute, pain; and one among the other advantages of the use of atropine in the treatment of iritis is that it stops the act of accommodation by temporarily paralysing the structures on which the function depends, and so averts the pain that would be caused by its exercise.

Demand for Acc.—The need for the exercise of Acc. increases in proportion to the proximity of objects, or, in other words, to the increased divergency of the rays proceeding from objects as they are placed nearer to the eye, so that these more divergent rays may still be brought to a focus on the retina. If the angular difference between a given remote and given near point be equal to the angular difference between some other two relatively remote and near points, the amount of Acc. exerted in the two cases will be alike, irrespective of the length of range over which the Acc.

may be exercised. Thus in fig. 70, as there is no angular difference between PAB and BAC, an eye in transferring vision from an object B to another object placed at C will employ the same amount of Acc. as in transferring vision from an object at infinite distance from which the parallel rays PP are derived, to another object at B, although the range is very different in the two instances indicated.

Mechanism of Acc.—The act of Acc., which in ordinary optical instruments is effected by altering the mechanical adjustments subordinate to the lenses, in the eye is effected by changes in the form of the crystalline lens itself. The distance of the surface on which the images are received, the retina, from the front of the refractive media, the cornea, remaining the same, it can only be by increasing the refractive power of the eye that the more divergent rays from nearer objects can still be brought to an exact focus at the distance of the plane of the retina. Various experimental observations have proved that the agent in affording this increase of refractive power is the crystalline lens; and, indeed, that this lens does become more convex, and so more refractive in exact proportion to the increased nearness of objects looked at by the eye, has been demonstrated by various investigators, and especially by Professor Helmholtz by means of his ophthalmometer. The anterior pole of the crystalline lens advances, while the diameter of its equatorial circumference is lessened, the whole anterior surface being thus rendered more convex and coming nearer to the cornea in proportion to the nearness of the objects to be observed. This dynamic increase of refraction gives way when objects farther off are to be observed; and when objects at a distance of 20 feet or beyond, from which rays proceed in nearly parallel lines, are looked at, the lens resumes the form which belongs to its state of repose. When no Acc. whatever is exerted, rays of light entering the eye, so far as regards the direction which they subsequently assume, are influenced by the stationary refractive quality of the eye alone; and although this permanent quality may be added to by the exercise of the function of Acc. as just described, there are no means provided within the eye by which it can be lessened.

The production of the changes in the form of the lens depends principally on the action of the ciliary muscle. The iris has ceased to have any power of action or has even been totally absent; the external ophthalmic muscles have all been paralysed; and yet the accommodatory faculty has remained unimpaired. This faculty cannot, therefore, be dependent on the influence of any of these structures as it was formerly supposed to be.

A further proof that the ciliary muscle is the agent, and the crystalline lens the anatomical structure acted upon by it, in the production of Acc., is the fact that whatever paralyses the ciliary muscle, by paralysing the motive power, stops the faculty of Acc., while the displacement or absence of the crystalline lens on which the ciliary muscle acts, equally arrests the faculty of Acc. The use of atropine, and conditions of disease interfering with the nervous supply to the ciliary muscle by the third pair of nerves, stop the faculty of Acc. It is equally arrested by removal of the crysta.line lens in the operation for cataract, or by its dislocation from accidental injury.

It is supposed by some that the circular fibres of the ciliary muscle, contracting in the manner of a sphincter, exert through the medium of the ciliary processes and the unyielding fluid in the canal of Petit a pressure on the equatorial circumference of the lens, and thus cause it to change its form for accommodation. Just as in the case of the iris, the circular fibres here referred to are antagonised by others disposed radially and forming the great bulk of the ciliary muscle. Indeed, but for the contractile reticulated structure filling up the space between the ciliary muscle and sclerotic (canal of Schlemm), the radiating fibres would be predominant. This will be at once apparent on referring to the vertical section of the forepart of the eye shown on the frontispiece. The meshes of the network are generally rhomboidal in form, with the long axis in the vertical direction, so that when their connections with the sclerotic are severed, they present the appearance of a circular muscle, and, indeed, have been described as such by my late colleague Inspector-General Dr. Macdonald, R.N. By the action of this structure, while the sclerotic is drawn inwards, a centripetal pressure is exerted upon the ciliary circle and muscle, which latter it would therefore antagonise, and so tend to increase the refractive power of the lens in the manner above explained. Though scarcely perceptible to us, the strain involved in the function of Acc. is very considerable, and amply accounts for the gradually increasing thickness of the hyaloid as it passes behind the capsule of the lens.

Power of Acc.—The *total power* of Acc. corresponds with the degree in which the function of Acc. can be exercised, or, in other words, with the amount of dynamic refraction available. The complete amount of power of Acc. is represented by the Acc employed when the near point of distinct vision is obtained. As the position of the near point of distinct vision depends on the accommodatory power being exerted to its fullest amount, this near point also furnishes an index to the limit of the power of Acc.

Expression of Power of Acc.—The power of Acc. exerted in order to obtain distinct vision at the nearest point at which an eye is capable of seeing objects clearly may be measured and expressed by the power of a converging lens, which, on being placed before the eye, will give clear vision at the same distance without the eye exerting any accommodatory power at all. If an eye be Emc. the stationary refractive power of its dioptric system is such that parallel rays of light proceeding from objects at infinite distance are brought to a focus on the retina, and while no Acc. is employed, it can only bring such rays to a focus at the plane of the retina as have a parallel direction with each other when they fall on the eye. It is evident, then, that a lens added to it which will represent the additional refractive power that the eye can obtain by bringing into play the whole of its Acc. must be such a lens as will cause the rays of light proceeding from objects at the distance of the nearest point of distinct vision of the eye, after passing through the lens, to impinge on the eye as parallel rays. The eye otherwise would not be able to bring the rays to a focus on the retina. The principal focus of this lens must therefore be at the same distance as that of the nearest point at which clear vision can be obtained. The rays starting from this near point, or, in other words, from the principal focus of the lens employed, will then be rendered parallel as they fall on the eye (see fig. 71).

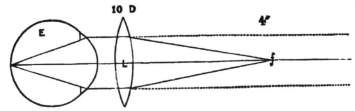

FIG. 71.—POWER OF ACCOMMODATION.

к. Emc. eye in repose and adapted for focussing parallel rays. The 10 D lens, L, represents the amount of Acc. which must be exerted to enable к to see an object at a distance of four inches.

If, for example, an Emc. eye can obtain, by the full exertion of its accommodatory power, clear vision of objects at a distance of 4 inches, or one-tenth of a metre from itself, then, without any Acc. being exerted a + 10 D lens, whose principal focus is at the same distance, will effect the same result. This lens, therefore, will represent and give the measure of the full amount of Acc. possessed by the eye under consideration. If the duodecimal or inch system of notation be employed, the power of the lens representing the measure of the Acc. exerted will be $\frac{1}{4}$; if the metrical system be employed, the power of the lens representing the measure of the Acc. will be +10 D.

Measure of Power of Acc.—It is constantly a matter of importance in ophthalmic examinations to determine and define the amount of power required for adapting V. from an object at some known distance to another object at some known nearness, and Professor Donders gave the following formula by which it can be readily found and expressed :—

$$\frac{1}{A} = \frac{1}{P} - \frac{1}{R}.$$

In this formula R is the measure of distance of the remote point for which the eye is primarily adjusted, whether expressed in inches or centimetres; P is the measure of the nearest or proximate point for which the eye becomes adjusted when Acc. is exerted; A is a convex lens, which, by its action on the divergent rays coming from the proximate point P, so lessens their divergency that they follow the same path, after passing through the lens, as was followed by the rays coming from the remote point R, without the lens; and $\frac{1}{A}$ is the power of this lens. It is obvious that the eye, in transferring vision from any remote point, R, to any near point, P, must so alter its focal adjustment as to act on the more divergent rays from P exactly in the manner attributed to the supposed lens A, and must exert a *power of accommodation* equivalent to $\frac{1}{A}$; $\frac{1}{A}$ therefore represents and expresses the difference in the Acc. required for vision at P as compared with that required for vision at R (see fig. 72).

Thus, to apply the formula to the example above given of an emmetropic eye adjusted for infinite distance in a state of rest, and capable of altering its focal adjustment by the exercise of Acc , so as to see clearly small objects as near as 4 inches: $\frac{1}{A} = \frac{1}{4} - \frac{1}{\infty} = \frac{1}{4}$, or the power of Acc. will be equal to $\frac{1}{4}$. In other words, the eye, in altering its adjustment from clear vision at infinite distance, when no accommodatory exertion was required, to vision at a distance of 4 inches, has exercised its accommodation, and has increased its refractive power to an extent equivalent to a 10 D lens, or a lens whose principal focal distance is at a distance of 4 inches from its centre. Thus when the position of the near point of distinct vision in an Emc. eye is found, the total amount of Acc. is also found. In the same eye, if vision be transferred from an object placed at 20 inches' distance to another object only 10 inches off, the formula shows that the amount of Acc. employed in the proceeding, or $\frac{1}{A}, = \frac{1}{10} - \frac{1}{20} = \frac{1}{20}$, or 2 D.

In a Mc. eye, whose farthest point of distinct vision in a state of rest is 12″, and whose nearest point of distinct vision by the exercise of Acc. is 3″, $\frac{1}{A} = \frac{1}{3} - \frac{1}{12} = \frac{1}{4}$, or the total power of Acc. is equal to 10 D. In other words, this eye, in altering its adjustment for vision at a distance of 12 inches to vision at the nearer distance of 3 inches, has increased its refractive power to an extent exactly equal to that of the Emc. eye quoted in the previous paragraph, which transferred its vision from an object at infinite distance to another at 4 inches' distance, viz. to an extent equivalent to that of a 10 D lens, or of a lens whose principal focal distance is 4 inches. In each case the power of accommodation exerted is = 10 D.

To determine the amount of Acc. for a Hc. eye, the total amount of H., or refractive deficiency, must be first found, and subsequently the position of the near point of distinct vision when H. is corrected. The lens which corrects H., added to the lens whose

principal focus is at the same distance as that of the near point,
gives the total amount of Acc.   Thus if H. = − 2 D, the correcting
lens will be a + 2 D lens, and if the near point when this lens
is in use be at 20 inches' distance, an equivalent + 2 D or + $\frac{1}{20}''$
lens must be added, and $\frac{1}{\lambda} = \frac{1}{20} + \frac{1}{20} = \frac{1}{10}$ or 4 D.   A + 4 D convex
lens or a lens of 10″ focus represents the Acc. necessary for ob-
taining a near point of distinct vision at 20 inches' distance for an
eye that is hypermetropic 2 D.

Under the metrical system, instead of the formula $\frac{1}{\lambda} = \frac{1}{p} - \frac{1}{\pi}$, the
formula Acc. = pD − rD will furnish the refractive power of the
equivalent lens (see fig. 72).   Thus in the examples above given, as
regards the emmetropic eye, in the first instance Acc. = 10 D − ∞
= 10 D, in the second instance, Acc. = 4 D − 2 D = 2 D ; as regards
the myopic eye, Acc. = 13·33 D − 3·33 D = 10 D ; as regards the
hypermetropic eye, Acc. = 2 D + 2 D = 4 D.

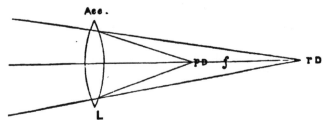

FIG. 72.—Measure of Power of Accommodation.

Acc. = pD − rD : f, the principal focus of the lens, L.

Influence of Varieties of Refraction on Acc.—The power of Acc.
is similar in all eyes that are healthy in condition, whatever may
be their refractive qualities, at corresponding periods of life.

Region of Acc.—The range, or latitude, or region of Acc. signifies the particular tract over which an eye can see clearly, extending from the most distant to the most near point of distinct vision ;
or, in other words, comprehends the space between the distant
point of distinct vision and the near point of distinct vision throughout which an eye can adjust itself so as to obtain clear images of
objects.   As shown in the preceding paragraph, an equal power of
Acc. may be exerted over regions which vary very greatly in regard
to position and extent of space.   In the first example a power of
Acc. equal to a 10 D lens is associated with a region of Acc. which
extends from infinite distance to a near point at 4 inches' distance from the eye.   In the second example the same power of Acc.
exists with a much more limited region of Acc., viz. one whose
limits are between a distant point of 12″ from the eye and a near
point of distinct vision at 3″ from the eye—that is, a region of 9
inches.

Although, as already stated, the *power* of Acc. is alike in all eyes of healthy condition at corresponding ages, the *region* of Acc. differs at corresponding ages in different eyes according as they are Emc., Mc., or Hc.

Association of the Internal Recti and Ciliary Muscles in Accommodation.—When an object situated in a plane midway between the two eyes, and at a short distance in front of them, is looked at, the two internal recti muscles, in giving the necessary direction to the two eyes for the visual lines to meet in the object, act concurrently with the ciliary muscles in their action for obtaining the amount of Acc. which is required to secure accurate vision of the object at the distance at which it is placed.   Clear and disembarrassed vision of a near object can only naturally be obtained when the functions of Convergence and Acc. are in due harmony.   There is no reason for believing that the internal straight muscles have any specific influence on the production of Acc., nor any influence beyond what results from their action in bringing the eyes into a favourable position for receiving the diverging rays proceeding from near objects, or, in other words, for effecting a right direction of the visual lines.

When the internal straight muscles are put into a condition of great tension for effecting a high degree of convergence, they may, perhaps, act as a *point d'appui*, as it were, on which to base greater efforts on the part of the ciliary apparatus, and in this way they may lead to an increased amount of Acc. and to clearer images being obtained of the near objects under view.   The general action, however, of the internal recti muscles, as regards Acc., is an associated not a controlling one.   The extent to which the two functions are employed in their associated action is liable to variations.   When the two eyes are directed to a small object at a fixed distance, and the object is seen clearly, under normal conditions the Convergence and the Acc. may be assumed to be in precisely due proportion. But when two prisms of small refracting angles are placed before the eyes in the manner described at p. 32 (see Prisms), the object may still be seen with precision, so that while the degree of Convergence is altered the Acc. remains unchanged.   On the other hand, while looking at the same object, if instead of the prisms weak + or − glasses within certain limits are placed before the eyes, the object may still be seen clearly, so that now, while the degree of convergence is unaltered, the amount of Acc. is changed.   The Acc. relatively to the convergence is, therefore, not a constant quantity, and the relations between the two functions are found to differ in different persons.   As, however, the convergence of the eyes increases in proportion as the object required to be seen is brought nearer to them, and the demand upon the accommodatory muscle must also increase at the same time, the actions of the internal recti and of the ciliary muscles are under normal conditions of necessity intimately and, for the most part, are almost proportion-

K

ately associated. It is essential that the facts of the intimate association of these two functions when they are free to act in concert, and of the difficulty of converging the eyes without the accommodatory force associated with the degree of convergence concerned being involuntarily called into action, should be fully recognised, for they explain many phenomena in visual diagnosis which without this knowledge would not be understood.

If, under any circumstances, the correlation between the action of the ciliary muscles in Acc. and the action of the interni recti in convergence, be very largely disturbed, a painful and almost intolerable sense of strain is experienced as the result. This has been illustrated experimentally by Mr. Brudenell Carter in a very simple manner. Let a young person with Emc. eyes put on convex spectacles of five dioptrics and attempt to read with them. The strain and tension of the eyes will soon become unbearable. He will be very nearly in the same condition as a person having M. of five dioptrics who is reading at his distant point of distinct vision. He will be converging to a point at 8 inches' distance, while there will be no demand for the exercise of Acc. In the myope, his amount of M. takes the place of Acc. at the distance named; the convex glasses replace it in the case of the emmetrope. On this Mr. Carter observes: ' It may be shown by a simple experiment that this strain is not due to the convergence effort *per se*, but solely to the fact that the convergence effort is made during repose of the Acc. If we combine with the convex lenses prisms with their bases inwards, of sufficient power to rest the convergence as well as the Acc., reading immediately becomes easy. If we then give the same person concave spectacles, such that he is able to overcome them by Acc., and bid him look at the horizon, the strain upon his eyes will again become unbearable. He will be exercising a good deal of his Acc. to overcome the glasses, but he will be exercising no convergence, because the distance of the object of vision requires that, for the sake of fusion, he should keep his visual lines parallel. We may relieve the strain instantly by placing before the concave lenses prisms with their bases outwards, which will call for a convergence effort in addition to the Acc. effort; so that we create strain by causing either function to be performed singly, and we relieve strain either by placing both at rest or by calling both into play.' (' On Defects of Vision, &c.,' Lond., 1877, p. 138.)

PRESBYOPIA.

Definition.—Vision of old age, or a condition of sight which, owing to contraction of the range of Acc. due to advance of years, does not permit very small objects to be accurately defined and distinguished at the proximity necessary for their proper recognition. The absolute ' near point ' of distinct vision is removed further than 8 inches from the eye, while rays from objects at relatively remote distances are still brought to a focus upon the retina, in the same way as they would be irrespective of the Pr.

Optical Conditions.—There is no change in simple Pr. as regards the stationary refractive quality of the eye, or, in other words, as to the eye in its state of repose ; it may be Emc., and neither concave nor convex lenses improve the sight for distant objects. But the necessary change of form of the crystalline lens cannot be obtained, or, in other words, the dynamic refractive power of the eye cannot be sufficiently exerted, for bringing the more divergent rays from very near objects to a focus on the retina.

Causes.—Natural changes from age. The near point of distinct V. recedes in all eyes progressively with increasing age ; the change is a normal one, and advances with advancing years. The recession of the near point commences early in life. The following table shows the decrease which observation has shown to take place in the power of Acc. at various periods between 10 and 75 years of age. The decrease of Acc. is expressed in dioptrics, and the recedence of the near point of distinct vision at the corresponding periods of life is shown in English inches.

Table showing the diminution in power of Acc., and the corresponding changes in the distance of the near point of distinct vision, as years advance.

Years	Dioptrics	Eng ish inches	Years	Dioptrics	Eng'ish inches
10	13·5	2·96	45	3 50	11·50
15	12	3·33	50	2·50	16
20	10	4	55	1·75	23
25	8·50	4·70	60	1	40
30	7	6	65	0·75	53
35	5·50	7 33	70	0·25	160
40	4·50	9	75	0	—

The change in Power of Acc., and the course of the proximate limit of distinct vision p.p. in the Emc. eye at different ages of life, is very plainly exhibited in the diagram on p. 132, copied from Donders (Ch. v. § 17), while it also represents the change in refractive power, and, consequently, in the far point of distinct vision, r.r., of the Emc. eye in advanced years. The diminution in the range, or full amount, of Acc. may be further recognised by observing the different distances at which P.P. and r.r. are separated from each other at different periods of life.

The recession of the near point of V. shown in the Table above is solely attributable to diminution in accommodatory power. As already explained. exertion of this faculty is called for in proportion to proximity of objects or divergence of rays. The gradual decrease of this faculty seems to be due to the crystalline lens becoming more and more firm with increasing years, so that the same

amount of alteration in its form cannot be effected. It is doubtful
whether changes in the structure and curvature of the cornea máy
not also assist in causing the difficulty of adapting the eye for near
vision.

From the explanation just given of the nature and causes of Pr.
it is evident that not only when an eye is Emc., or Hc., but also if
it be Mc.. it may equally become subject to Pr. from age. Whenever the actual near point of distinct vision becomes removed
further than 8″ from the eye, Pr. has arrived.

Table exhibiting the changes in accommodation in an emmetropic eye at different periods of life.

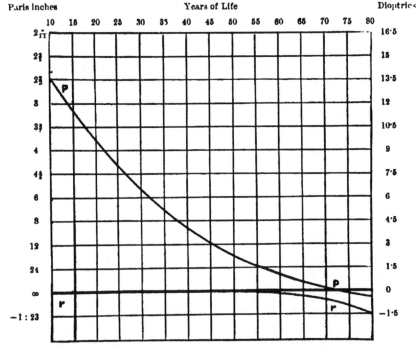

The slde figures show the amount of Acc. expressed in French inches and in Dloptries.
P P shows the cours of the proximate, r r of the remote, joint of distinct vision. The
distance between the two lines P P and r r indicates the range of Acc.

Abnormal or False Pr.—Diseased conditions, such as interference
with the elasticity of the lens capsule by posterior synechiæ after iritis,
morbid changes in the structure of the capsule of the crystalline
lens or lens itself, paralysis, complete or partial, of the ciliary muscle,
and any diseases inducing great constitutional debility, may increase
the presbyopic condition temporarily or induce symptoms similar
to those of Pr. Presbyopic vision resulting from these causes should
be regarded as abnormal and distinguished from True Presbyopia,
the result or normal changes from advancing years or old age.

Symptoms.—Normal Pr. is not usually complained of by persons with Emc. vision till after 40 years of age, generally not till 45 years. The degree of complaint will vary according to the size of the objects the subject of Pr. habitually deals with. The presbyope says that, though he can see as well as formerly at a distance, he finds it troublesome, or even painful, to try and recognise small objects near him. In reading he has to remove the print farther and farther from his eye. Although he can for a short time, by much exertion, read at a distance of 8 or 10 inches from the eye, the letters soon become misty, his eyes ache, and he is compelled to remove the print farther away. Then, on account of the distance he has to hold the print, he cannot see the words accurately.

The difficulty is greater in the evening with artificial light than during the daytime with sunlight, and he sees best when he holds a candle or lamp close to the page of the book he is looking at.

The deterioration that takes place as age advances in the transparency of the principal dioptric media which is perfect in healthy eyes in the earlier period of life, and the haziness with more or less tinge of colour in them when age is far advanced, prevent ordinary sources of light from giving the full amount of contrast between the white and black parts of a printed page or between the lights and shadows of other objects, and thus prevent the images of the objects themselves from being impressed on the retina with the same vividness as in former earlier years of life. This condition, together with the greater distances at which small objects have to be held, and consequently the lessened amount of light reflected from them at such distances, combine to explain the difficulty old persons meet with in reading or doing fine work without the assistance of a strong light upon the objects in view. One advantage in the correction of Pr. is that the need for such a strong light is considerably lessened, because, after correction, the print or other small objects can be brought near to the eyes and yet sufficiently recognised.

The nearest point of distinct vision can only be attained by extreme exertion of the ciliary apparatus, and, when this state of tension is produced, it quickly causes fatigue. It is to obviate this annoyance that objects are removed further off than the actual near point, but then, if the objects are very small, their retinal images become too minute, and the light emitted by them too diffuse and too diminished for easy recognition. The objects are again brought nearer to the eye, the ciliary strain is again induced, and this course, on being persevered in, gives rise to ocular aching, super-ciliary pain, headache, and vertigo, until at last a stop is put to the power of doing the fine description of work in which the patient may have been occupied.

Diagnosis.—The age of the patient, with the usual evidences of the adjustment of the eyes for parallel rays if the patient be

Emc., as explained under Em., and the distance of the near point, sufficiently distinguish Pr. from any other affection. When either M. or H. co-exists with Pr., there will be the evidences already described of these conditions in addition to the fact of the near point of distinct vision being distant beyond 8″ from the eye.

To Determine the Degree of Pr.—As Pr. is generally held, in accordance with the limit laid down by Professor Donders, to exist when the near point under binocular vision has receded beyond 8″ from the eye, the degree of Pr. is determined by finding the focal power of the convex lens which will bring back the near point to 8″.

The amount of Pr. is the amount of the accommodatory power which has been lost from that which previously enabled the person to work continuously at fine work at 8″ distance from the eye without strain or fatigue, relatively to the remainder which now only admits of clear vision at some distance beyond 8″. The convex lens which will so add to the refractive power of the eye as to bring back the near point to 8″ substitutes itself for the accommodatory power which has disappeared from the eye itself, and therefore serves as an expression for the missing power, or, in other words, for the degree of presbyopia.

The refracting value of the required lens is found by using the formula which has been already explained as serving for an expression of power or latitude of accommodation, viz., $\frac{1}{A} = \frac{1}{P} - \frac{1}{R}$, or the formula, Acc. $= pD - rD$. As $\frac{1}{A}$, or Acc. thus determined, represents the missing latitude of Acc., Pr. may be substituted for either ; so that Pr. $= \frac{1}{P} - \frac{1}{R}$, or $pD - rD$. The proximate point P, or pD, is here 8″ or 5 D, because that is the distance of the visual near point for which a substitute for the lost Acc. is required, and to which the refractive power of the eye is to be adjusted.

The remote point, R, is the distance to which the natural near point has receded owing to Pr., or the distance up to which Acc. is missing.

Examples.—Suppose the near point has receded to 12″ distance from the eye; then Pr.$=\frac{1}{8} - \frac{1}{12} = \frac{1}{24}$, or 5 D$-3\cdot34$ D$= 1\cdot66$ D (24·09″ Eng.). If it has receded to 24″ distance, then Pr.$=\frac{1}{8} - \frac{1}{24} = \frac{1}{12}$, or 5 D$-1\cdot66$ D$=3\cdot34$ D (11·97″ Eng.).

To Determine the Correcting Lens.—The same formula that determines the amount of accommodatory power which is missing, determines the focal power of the lens which must be added to the eye to act as a substitute for it. Thus in the example first given, in which the Pr.$=\frac{1}{24}$, or 1·66 D a + 24″ lens, or a lens of 1·66 D, will neutralise the Pr. ; in the second, a + 12″ or 3·34 D lens. With the aid of these lenses, viz. convex 1·66 D and 3·34 D, the presbyopic patients will be able to do fine work at 8″ distance from the eye as if the near point of distinct vision of the eye itself were at that distance.

Power and Range of Accommodation in Presbyopia.—The power and range of Acc. of the presbyopic eye diminish in proportion as the Pr. increases, or, in other words, in proportion as the Acc. for near objects is lost. Thus in the Emc. person whose remote point of distinct vision is at infinite distance, at about 10 years of age the power of Acc. is $\frac{1}{F.85}-\frac{1}{8}=\frac{1}{r.25}$, or 14 D; at 20 years is $\frac{1}{4}-\frac{1}{8}=\frac{1}{4}$, or 10 D; at 30 years $\frac{1}{6}-\frac{1}{8}=\frac{1}{4}$, or 7 D; at 40 years $\frac{1}{8}-\frac{1}{8}=\frac{1}{9}$, or 4·50 D; at 50 years $\frac{1}{16}-\frac{1}{8}$, or $\frac{1}{16}$, or 2·50 D; at 60 is $\frac{1}{40}-\frac{1}{9}$ or $\frac{1}{40}$, or 1 D; and at 75 is 0. The power of Acc. is thus shown to be considerably lessened as age advances. While the boy of 10 years of age can increase the refractive power of his eye to the amount of 14 dioptrics, the man of 30 can only increase it to about 7, of 40 to 4·5 dioptrics, while at 75 the power of adding to the refractive power of the eye, or the accommodatory power, has disappeared altogether.

Presbyopia with M.—As the power of Acc. is alike in all eyes at corresponding ages, and as the distant point of distinct vision is nearer to the Mc. eye than it is to the Emc. eye, it follows that the near point of distinct vision will be nearer to the Mc. eye than it is to the Emc. eye at corresponding ages of life. The near point will not recede beyond 8″ so soon in the Mc. person as in the Emc. person, or, in other words, Pr. will appear later in life in the Mc. person than in the person with Emc. vision. Moreover, the Pr. can only occur in Mc. persons whose M. is moderate in degree. If the M. amount to $\frac{1}{8}$th, or 5 dioptrics, Pr. can never occur; for the distant point of distinct vision being at 8″ distance from the eye, so long as any power of Acc. remains, the near point cannot be removed so far as a distance of 8″ from the eye; and even when all accommodatory power disappears or is lost, whether naturally or by atropia paralysis, the near and distant point of distinct vision will merely coincide at a distance of 8 inches from the eye. Still less can a person whose Mc. is higher than $\frac{1}{8}$th, or 5 D, become presbyopic.

If the M. be moderate in amount, then, as age advances, the near point may recede beyond 8″; and in such a case the M. will require to be corrected by concave lenses to enable the person to see distant objects clearly, while at the same time the Pr. will have to be corrected by convex lenses to make up for the deficiency of Acc. for seeing small objects at near distances.

Presbyopia with H.—As in the Hc. eye the distant point of distinct vision is negative, and infinite rays require a certain amount of convergence to be imparted to them, in order that they may be brought to a focus on the retina the Acc. is trenched upon for affording this required convergence. The H. is thus not immediately apparent; it is covered by the Acc. which it borrows, as it were, from the reserve for near objects, to use it for obtaining clear vision of distant objects. The H. thus becomes concealed or is *latent*. If the H. is very moderate in amount, and the subject is young so that Acc. is in abundance, the existence of H. may readily be overlooked. The amount of Acc. which is borrowed for the service of the H. in supplying the refractive deficiency has, under such circumstances, very little effect on the position of the near point in respect to change of its working distance, and this change may readily therefore escape notice. If a young person who is Emc. can exert Acc. equivalent to a convex lens of 12 D the near point of distinct vision will be $3\frac{1}{3}$ inches from the eye; if the Acc. be lessened by 1 D the near point will only recede to about $3\frac{2}{3}$ inches; and if lessened by 2 D, to 4 inches—differences which would be too slight to be noticed. If the Acc. were completely removed, as by atropia paralysis of the ciliary muscle, the H. would be readily detected, for a suitable convex lens would improve distant vision. But without this being done the H. may be unnoticed until a more advanced period of life.

If, however, the H. be considerable in amount, the early recedence of the near point of distinct vision, or what is tantamount to an early occurrence of Pr., will sufficiently prove its existence. As the distant point of distinct vision in the hypermetrope is, as it were, further from the eye than it is in the Emc. eye, while the power of Acc. is the same at the same ages in both, it follows that, under all ordinary conditions, the near point of distinct vision will be further removed from the Hc. eye than it is from the Emc. eye at corresponding ages of life. If the H. be excessive, the near point will be beyond 8″ very much sooner in the Hc. person than in the Emc. person, and Pr. will occur proportionally earlier in life in such a hypermetrope than it does in the emmetrope. This early occurrence of Pr. at once proves its association with H. When Pr. is fully established in an adult hypermetrope, his H. will require to be corrected by a suitable convex lens to secure clear vision of distant objects, while his Pr. will have to be corrected by a lens of still greater converging power—one that will not only correct the general deficiency of refractive power of the eye, or the H., but that will also make up for the loss of accommodatory power so as to enable the eye to deal without inconvenience with small objects at a near distance from the eye.

CHAPTER V.

On Impaired Vision connected with Strabismus—Strabismus—Definition—Forms of Strabismus—Directions of Deviation—Apparent Strabismus—Images in Convergent and Divergent Strabismus—Concomitant Strabismus—Diplopia—Diagnosis of Concomitant and Paralytic Strabismus—Measurement of Deviation in Strabismus—Treatment—Influence of Strabismus as regards Military Service.

STRABISMUS.

Definition.—Squint. Abnormal obliquity of the optic axis in one or both eyes. Malposition of the two eyes relatively to each other of such a kind that the directions of their optic axes are not in mutual accord. The visual lines do not meet together at the distance at which a given object is looked at, and true binocular vision is consequently unattainable by persons affected with strabismus.

Various Forms of Strabismus.—Two distinct forms of strabismus, due to different causes, are met with ; one functional, depending directly upon irregular muscular action, originating in, or accompanying, anomalous states of refraction, or impaired vision of one or both eyes, and generally distinguished by the name of ' concomitant strabismus,' from the squinting eye accompanying the normal eye in all its movements ; the other pathological, produced by partial or complete paralysis of one or more external muscles of the eyeball, or due to pressure from causes existing within the orbit. In the first of these two kinds of strabismus all the muscles act, but the normal concert of action is not maintained in the binocular movements ; in the second, the movements of one or more of the muscles are impaired or entirely arrested. The term ' strabismus ' is employed by many surgeons to designate only the former kind, or true concomitant strabismus; absence of concordance in the visual axes from defective innervation or complete paralysis being classed under other special headings, such as paralysis of the ocular muscle or muscles affected, or under diplopia.

Directions of Deviation in Strabismus.—Strabismus is commonly either *convergent*, the squinting eye being turned toward the median line ; or *divergent*, being turned outwards. The deviation may, however, be upwards or downwards, though it is rarely so. In convergent and divergent strabismus the retinal images of an object are separated laterally, but remain in the same horizontal plane ; in the other directions of deviation, they vary in altitude and inclination according as the superior and inferior recti or oblique muscles are involved.

When the subject of convergent strabismus looks at an object one eye only is visually directed upon it, the other eye has its visual line directed inwards; in the same way, when the subject of divergent strabismus looks at an object, one eye only is directed toward it, but the other eye has its visual line directed outwards. In either case, while the sight is thus fixed upon the object the deviation is limited to the eye which is turned inwards or outwards.

The deviation may only occur at intervals, or after occasions of long-continued employment of the eyes at near objects, and under conditions of unusual ocular strain and fatigue, or general weakness; but although the occurrence of the deviation, or of the strabismus, may thus vary, the direction in which the deviation takes place, when it does occur, will always be alike in the same eye.

Apparent Strabismus.— A condition which simulates strabismus is sometimes noticeable, and is usually referred to under the name of *apparent strabismus*. It depends upon the angle formed by the meeting of the visual line with the line of the optic axis at their point of intersection within the eye. (See fig. 5, p. 5.) The relative positions of the two eyes, as elsewhere mentioned, are judged of by observing the relative positions of the centres of the two corneæ with respect to each other and to the middle points of the palpebral openings. When two Emc. eyes are regarding a distant object, the visual lines of the eyes are practically parallel, and as these lines pass a little internal to the centres of the corneæ, it follows that the centres of the corneæ must be turned a little outwards in respect to the middle points of the palpebral apertures. In Emc. eyes this divergence, or apparent strabismus, is so slight as to be hardly observable even on close inspection. The divergence is, however, increased in Hc. eyes, as the angle formed by the meeting of the optic axis and visual line is greater, and the apparent strabismus becomes all the more marked. In myopic eyes this angle is lessened, and the visual line may even change position and pass outside of the optic axis, when the appearance of strabismus may be presented, but with a convergent instead of a divergent direction.

Images of Objects in Convergent and Divergent Strabismus.—In *convergent* strabismus the two images of an object are formed on the corresponding retinæ; that which appears on the right side belongs to the right retina, that on the left side belongs to the left retina. In *divergent* strabismus the images are crossed and formed on the opposite retinæ; that which appears on the right side belongs to the left retina, that on the left side to the right retina. The relative positions of the images of an object in some instances of diplopia depending on slight degrees of muscular inefficiency, in which the deviation is so limited as to be hardly noticeable by ordinary observation, are almost the only means by which a surgeon is enabled to decide what particular muscle or muscles are affected. Their respective situations in relation to the two eyes will be rendered obvious to the patient at once if the two images are seen by him under different colours, as may be effected by causing the patient to look with one eye through stained glass (see figs. 3 and 4). If the deviation be horizontal, and the images are homonymous, there is convergent displacement, and the external rectus muscle is affected; if crossed, there is divergent displacement, and the internal rectus muscle is affected.

The frequent associations of convergent strabismus with hypermetropia, and of divergent strabismus with myopia, have been already remarked on in the sections on these refractive conditions.

Concomitant Strabismus.—Concomitant strabismus may be only occasional or it may be persistent. The deviation may alternate between the two eyes, each in turn being capable of directing its visual line on the object while the other deviates; or one, or both eyes, may be permanently displaced. When the deviation alternates between the two eyes, the amount of deviation in the right eye when the visual line of the left eye is directed upon the object looked at will be equal to that of the deviation in the left eye, when the macula lutea of the right eye is in line with the object.

In the strabismus of muscular paralysis the loss of power of the eye to move in the line of impaired, or lost, action of the paralysed muscles and the deviation in the opposite direction are constant.

When the strabismus exists only to a moderate amount, and the deviation is limited to one eye, it is sometimes not easy of detection so long as the two eyes are observed together. But it may usually be readily demonstrated, if the retinal functions of the two eyes are preserved, in the following way. The person under observation is directed to regard an object at the distance of a few feet in front of him. Each eye is then alternately covered by the hand of the observer. When the eye whose visual line is directed to the object is covered, the squinting eye may be noticed at once to change its position. It will assume a direction which will bring it visually in line with the object; and equally, on again uncovering the other eye, it will recede from the new to its former place.

Diplopia.—Whenever the physiological harmony of the visual lines is interfered with in binocular vision, as happens with strabismus, the discord necessarily causes the formation of double retinal images, but does not necessarily cause diplopia, or double vision.

Diplopia does not exist in the large majority of cases of concomitant strabismus, but is constantly found in strabismus due to paralysis, or partial inefficiency of one or more of the ocular muscles.

There are several reasons why the subject of concomitant strabismus can avoid taking note of the image in the displaced eye, while the subject of paralytic strabismus is unable to do so. In the former, the eye whose visual line is properly directed on the object has its image formed on the macula lutea, and it is of course relatively sharp and distinct; while the image in the displaced eye is pictured on a part of the retina away from the macula lutea, and is, therefore, more or less dull and indistinct. The relative displacement of the two eyes, or the strabismal angle, remains constant, for in concomitant strabismus there is no interference with the range of the muscular movements of the globe of the eye, as in paralytic strabismus, but only with the *position* of the range over which the movements take place. In consequence, the indistinct image on the retina of the displaced eye is always formed at a fixed part of the retina; just as much so as the sharp image is formed on the macula lutea of the normally directed eye. The retinal sensitiveness of the part where the relatively dull or spectral image occurs becomes lessened at first from exhaustion and subsequently from disuse, and the suppression of mental attention to the image by the subject of concomitant strabismus in the efforts which he instinctively makes to rid himself of the troubles of double vision, becomes easily established. On the other hand, in strabismus due to paralysis of one or more muscles, the relative positions of the two retinæ are constantly changing as objects in different positions are looked at, and corresponding changes constantly occur in the situations of the retinal images. If, for example, the external rectus of one eye be paralysed, causing a certain amount of inward deviation of this eye, as the sound eye turns inwards, or in a direction toward the paralysed muscle of the deviated eye, the strabismus is rendered more and more obvious—the strabismal angle becomes greater, and the distance of separation of the double images in the two eyes is increased. If the sound eye turn outwards, or in a direction away from the paralysed muscle of the deviated eye, the two eyes in the case supposed will assume a direction more in concert, and at a certain point the visual lines may even both centre in the object, and the diplopia disappear. The strabismus is thus seen not to be constant, as it is in concomitant strabismus, but varies according to the position of the object looked at, and consequently the position of the image in the displaced eye is equally shifting and inconstant. There is little to interfere with the retina retaining its sensitiveness over its whole area, and under these circumstances the strabismal subject finds himself unable to suppress perception of the second image when it ever occurs.

When concomitant strabismus happens in very early life, the suppression of the images in the squinting eye seems to take place with the utmost facility, amblyopia is established, and if the strabismus be not rectified, the retinal sensitiveness becomes lost beyond recovery.

Diplopia, due to slight muscular paralysis, may be noticed by a patient before strabismus is visible to an observer. It may amount to little more than mere haziness from overlapping of the two images at varying distances, but may still be accompanied by a difficulty of co-ordinating the touch with the sight of objects, as well as by the vertigo and other subjective symptoms to which diplopia usually leads. If two separate images are noticed, the sharper image produced on the macula lutea of the sound eye, and the weaker image on the retina of the deviated eye, may be distinguished by the aid of the colour test elsewhere mentioned, if they are not otherwise recognisable; and the recognition as soon as established will indicate not only which eye is disordered, but also the affected muscle or muscles. If two images of a bright object, as of a lighted candle, placed at a distance, are seen, and one image is coloured by the intervention of coloured glass before one of the patient's eyes, the patient will readily point out to the observer the situation of the image belonging to the eye before which the coloured screen is placed and that belonging to the other eye. The observer will at once see, from the relative positions of the two images, in which of the patient's eyes the true image is formed. The affected eye is thus made known, and the situation of the displaced image, relatively to that of the true image, whether it be to its right or left hand, or above or below, indicates the muscle or muscles which are at fault in this eye. The displaced image always appears on the side opposite to the ocular deviation, so that if the images are homonymous, convergent strabismus is shown to exist, while, if they are crossed, divergent strabismus is indicated. If, for example, it be the muscle concerned in convergence which is affected, say of the right eye, leading to divergent displacement, the spectral image will be crossed to the left hand of the true image as seen by the patient; if it be the muscle of divergence which is affected, leading to convergent displacement, the false image will be to the right side of the true image as seen by the patient and indicated by him to the observer. (See figs. 3 and 4, p. 4.)

Diagnosis of Concomitant and Paralytic Strabismus.—The conditions already mentioned form sufficient means of distinguishing between these two forms of strabismus, but one other difference between them, which is useful for diagnosis, may be presently mentioned. The characteristics of concomitant strabismus, which have been already referred to, are: 1. The affected eye retains its full natural range of movement, but this range is displaced. Its movements do not take place in the normal arc, the eye turning more in one direction, and less in the opposite direction, than the normal eye; 2. The deviation of the affected eye, and the amount of disagreement in direction of the visual lines of the two eyes, at whatever distance the object may be placed which is looked at by them, are constant; and 3. The displaced, or spectral image, formed in the affected eye is mentally ignored, so that diplopia does not occur. On the other hand, in paralytic strabismus, 1. The range of movement of the affected eye is always lessened in the direction of the paralysed muscle; 2. The amount of deviation of the affected eye, and of dissociation of the visual lines of the two eyes, are not constant, but vary with the different directions of the objects toward which the patient turns his eyes; and 3. Diplopia always exists in the range proper to the paralysed muscle, in all cases in which the retinæ retain their sensibility.

The other symptom by which paralytic may be distinguished from concomitant strabismus is the following. If, while the head remains fixed in position, the healthy eye be covered by a diaphragm of frosted glass, at a time when the patient is looking toward an object, the affected eye will make a slight move, in order to try and adjust its visual line upon the object. This movement of the affected eye may be seen to be accompanied by a movement of the healthy covered eye, so that if the other is directed on the object, the covered eye becomes the squinting eye. In paralytic strabismus, this movement of the covered sound eye exceeds in extent the movement of the affected eye, because the slight movement of the affected eye, owing to its defective innervation, requires a greater effort of the will to effect it, than it would if the innervation were perfect. This increased effort affects the associated muscle of the sound eye, and so leads to a greater amount of movement through its agency. The deviation of the affected eye is usually described as the *primary* deviation; the movement of the sound eye under the circumstances named, as the *secondary* deviation. On the other hand, in *concomitant strabismus*, under similar circumstances, there is no difference in the extent of movement of the two eyes. When the healthy eye is covered, whatever may be the distance to which the affected eye moves in order to direct its visual line upon the object, the extent of the associated movement of the covered eye will be precisely the same. The rule is thus deduced that in all cases of paralytic strabismus the *secondary* deviation exceeds the *primary* deviation, while in concomitant strabismus the two deviations are exactly equal.

Measure of Deviation in Strabismus.—The linear measurement of the deviation of a squinting eye may be made in either of the following ways. Let the patient be directed to look at a distant stationary object with both eyes free, and while he is doing so, mark with a dot of ink on the margin of the lower eyelid of the *squinting* eye, the spot which would fall within a vertical line prolonged from the centre of the deviated cornea. Then close the normal eye, and let the affected eye be directed to the object, and, as before, mark a spot on the lower lid corresponding with a vertical line drawn through the centre of the cornea in its new position. The distance measured between the first and second spot gives the linear measure of the deviation. Or, while the two eyes are open, a spot is marked on the margin of the lower lid of

the *sound* eye in line with the centre of the cornea. A spot is then
marked on the border of the lower lid of the affected eye, which
exactly corresponds in position with that of the spot placed opposite
to the centre of the cornea of the sound eye. A second spot is then
made to mark the situation of the centre of the cornea as it exists
in the squinting eye. The distance which separates these two
spots indicates the amount of displacement of the optic axis. The
distance between the two spots can be measured by an ordinary
tape measure, graduated with sufficient fineness, but still more conveniently by means of a strabismometer. (See p. 49.)

Treatment of Strabismus.—Although in comparatively mild cases
of paralytic strabismus, such as sometimes result from constitutional states—from syphilitic, rheumatic, or diphtheritic disease,
for example—relief may often be effected by suitable remedies;
and also in the milder forms of concomitant, or functional, strabismus, by the aid of prisms, in exercising and gradually strengthening the weaker muscles, and so enabling them to overcome the
stronger action of their opponents, together with correction of any
abnormalities of refraction that may co-exist by appropriate lenses;
still, in the majority of cases, especially when the deviation is fully
established, it only admits of cure, or palliation, by the operation of
tenotomy. The proper treatment of strabismus usually requires
very close and accurate observation of each particular case concerned,
and a consideration of the subject does not fall within the scope of
this manual; a description of the various proceedings involved in
it will be found in all systematic works on diseases of the eye and
their treatment.

Influence of Strabismus as regards Military Service.—The disfigurement caused by well-marked strabismus, and the uncertainty
it gives rise to, as to the precise direction in which a man is looking, are sufficient reasons for rejecting a recruit seeking enlistment,
but the most important objections are the monocular vision on the
one hand, and the amblyopia on the other, with which concomitant
squint is usually accompanied. Operative interference may remove the disfigurement, but will not repair the amblyopia if it be
of long standing. I have known a candidate for an army commission get the concomitant strabismus with which he had been
affected from childhood, rectified by surgical operation, but a year
afterwards, the amblyopia of the eye which had been displaced
remained without any apparent improvement. Practically the
influence of concomitant strabismus on aptitude for military service
is settled by the ordinary tests for acuteness of vision. If the
soldier can count the test-dots *by each eye* in succession at the required distance, although a slight amount of functional strabismus
may be apparent, there is no sufficient cause for rejection: if the
test-dots cannot be counted at the regulated distance, whether the
impediment may be owing to the refractive condition of the eye, or

to amblyopia the result of disuse, the results as regards the unfitness of the candidate for military service in the ranks will be the same. As regards a candidate for an officer's commission, there will be this difference ; if the difficulty in the way of reading particular test-types be due to ametropia only, he may overcome it by wearing suitable glasses ; if amblyopia from disuse exist in addition, no glasses will rectify it so as to enable him to pass the required tests.

Paralytic strabismus is a sufficient cause for rejection of a recruit. It is therefore only likely to be met with after a soldier has been serving in the army, and, when once it has occurred, it is usually a cause of the man being discharged from further service, owing to its liability to recur, even though it may appear to be only temporary in its nature. In occasional cases, where it has been traceable to the effects of a slight attack of insolation in a tropical climate, it may disappear altogether on removal to a temperate station ; but under many other conditions, when induced by morbid states, as syphilis and others, though it may be removed for a time by appropriate treatment, there is no security against the risk of its return. When paralytic strabismus is fully established, the attendant diplopia, and loss of ocular movement in the direction of the paralysed muscle, and with it frequently loss of part of the field of vision, as a matter of course, incapacitate a soldier for further military service.

CHAPTER VI.

On Impaired Vision due to Defects of Colour-sense —Colour-blindness—Definition — Symptoms —Varieties —Causes of Colour-blindness—Acuteness of Colour-sense - Diagnosis—Holmgren's Test - Proofs of Kind and Extent of Colour-blindness—Thomson's Application of Holmgren's Test—Maréchal's Lantern-test—Colour-sight in the Royal Navy—In the Indian Government Services—Influence of Colour-blindness in the British Military Service—Amblyopic Colour-blindness.

COLOUR-BLINDNESS.

Definition.—Imperfection, or total deficiency, of visual perception of one or more colours. The term *dyschromotopsia* is employed to express dulness of the colour-sense, or a difficulty in distinguishing colours and their various shades and combinations; *achromotopsia*, to express complete absence of perception of one or more colours, or, in other words, total inability to distinguish truly one kind of colour from another. As a rule, both eyes are equally affected, but cases have been recorded in which monocular achromotopsia existed while the other eye was entirely free from the defect.

Colour-blindness is of two kinds, viz. (*a*) *congenital colour-blindness*, a defective state of vision, often hereditary, under which the subject, though possessing distinct perceptions of form, and of contrasts of light and shadow, and, indeed, normal sight in all other respects, is unable properly to recognise diversity of colour, or to perceive certain colours; and (*b*) *amblyopic colour-blindness*, deficiency, perversion, or loss of colour-sense associated with morbid states of the retina which are mostly manifest to observation.

Symptoms.—Colour-blindness may exist without the subject of it being aware of his defect, until his attention is directed to it by experimental trials, or until the occurrence of some observation or of some special difficulties consequent on his disability reveals its presence. In the case of one colour-blind person known to me, the existence of the defect was only discovered by his expressing surprise to his daughter that certain posts, which were partly designed for attracting attention, had not been painted of a different colour from the neighbouring shrubs. The fact was the posts had been painted red to distinguish them from the green objects near them. He was then of mature age, yet the existence of his colour-blindness had not been previously known to himself or others. On looking at the colours of the spectrum, some of them may be seen correctly, while others that are really very unlike, produce on the colour-blind subject a similar optical effect, and are only distinguished by appearing to vary in intensity of light and shade. The eyes may be normal as regards their refractive qualities and power of accommodation, visual acuteness, and appreciation of form, in short, perfect in all respects excepting in the colour-sense. The subjects of colour-blindness make great mistakes in sorting colours. Inability to distinguish red from green is the most common form of colour-blindness. One colour is regarded as similar to the other, only varying perhaps in shade or saturation. The red and green colours are confounded with grey as well as with each other, and, in consequence, are wrongly applied or distinguished. To persons who are 'blind for red,' objects coloured red appear to be of a darker hue than they do to those who are not colour-blind, apparently dark grey, at the same time that some reds are not distinguished from green; and similarly, the 'blind for green' see green objects darker than they really are, while some greens are confused with purple, blue, or grey. Such differences in objects as depend upon varieties of colour for their recognition are, therefore, of no avail to the colour-blind.

Varieties.—Instances have been recorded in which no distinctions of colour have been recognised; in which all objects have seemed to be uniform in colour, and simply varying in degrees of light and shadow, or depth of tint, as in the photograph of a landscape. Such instances have been very rarely met with. In ordinary achromotopsia, there is usually absence of perception of one, or at the most two, of the three colours, red, green, or violet. The subject may be blind for red, or for green, or for both red and green, or he may find a difficulty merely in distinguishing between red and green. Or he may be 'blind for blue' or violet, while he recognises red and green. This last variety appears to exist the least frequently of all the kinds of colour-blindness; indeed, to be very rare.

Colour-blindness should not be confounded with inability to name colours, nor even with dyschromotopsia—that is, inability to appreciate with facility or nicety the more complex colours or to see the colours of objects under the same conditions of extent, distance, or saturation, under which other persons can see them, or to distinguish readily different shades or tints of one and the same colour. These imperfections may, or may not, be due to a limited defect of organisation in the nerve-elements; usually they are simply attributable to want of sufficient education of the colour-sense, to enable the distinctions named to be recognised or appreciated. Just as the senses of hearing and taste may be developed by cultivation, so may the colour-sense when it exists. But in true colour-blindness, achromotopsia, the sense for certain colours does not appear to exist, and all attempts at improvement by education have consequently hitherto been of no avail. It is only this form of optical defect that, as a general rule, is of practical importance as regards occupations in civil life or in the military or naval services.

Causes of Colour-blindness.—The abnormal anatomical conditions on which colour-blindness depends have not yet been demonstrated, and even the true nature of colour-sensation is as yet undetermined. The theory most generally accepted for explaining perception of colour is that known as the Young-Helmholtz theory. According to this theory there are three distinct elements in the retina which are concerned in receiving visual impressions, one of which is chiefly excited by red rays, the second by green, and the third by blue or violet rays. If the elementary fibres which are concerned in the perception of these three fundamental colours are equally excited together in normal eyes, the resulting impression is that of white; while the impression of compound or intermediary colours depends on the fact of either two, or of all three, of the retinal colour elements being severally excited in greater or less proportion among themselves. The cause of defective or of completely deficient perception of colour depends upon a faulty or undeveloped condition, or upon a total absence of one or more of these retinal colour elements. Although the hypothesis just named is the one which is very generally accepted instead of the Newtonian view of the three primary colours being red, yellow and blue, and green and violet being mixed or compound colours, some philosophers are opposed to it, and there are certain undoubted difficulties in the Young-Helmholtz explanation which still require solution.

No morbid condition of the media or fundus of the eye can be detected by the ophthalmoscope in colour-blindness, a fact which distinguishes congenital achromotopsia from most cases of amblyopic achromotopsia.

Acuteness of Colour-sense.—The sense of colour, like the perception of shape and size, of objects, is most acute in the region of the macula lutea. It decreases in proportion as the images of coloured objects fall on the retina more and more remotely from this point, but the decrease is modified by the clearness of colour and degrees of illumination of these objects. The field of colour-vision is not alike for all colours, as may be readily observed by using test-objects of different colours with the perimeter. The perception of green is the most limited, that of red is next, while the visual field of blue is usually the most extensive. A white test-object can be seen beyond the limits at which one of a blue colour can be distinguished. The field of vision for each colour becomes restricted in the progress of certain morbid states of the retina.

Diagnosis.—Snellen's coloured types and other similar objects were formerly supposed to afford satisfactory means of testing achromotopsia. Five colours were supplied in Snellen's book of test-types as distinguishing tests for colour-sense, viz. red, yellow, green, blue, and grey. The orange, indigo, and violet colours of the spectrum were not included. In like manner small flags and lights of different colours were used for testing perception of different colours in medical examinations for the Royal Navy. But special attention has been given to this subject during the last few years, and the tests just mentioned are no longer regarded as reliable. They may still be employed for measuring the acuteness of central vision for coloured objects relatively to the visual acuteness for uncoloured objects, but not for determining whether the distinctions between the colours themselves are normally perceived. Professor Holmgren has proved that colour-blind persons have acquired in some instances a power to distinguish signals and lights of different colours, and even of calling them by their proper names, from the retinal impressions of different degrees of light and shade made by them—that is, so long as the signals and lights under observation are put before them under conditions of equal clearness. The colour-blind person has trained himself to observe these qualities as substitutes for his want of colour-sense; just as the totally blind educate and refine their senses of touch and hearing to make up for their loss of visual power. There have been repeated proofs that colour-blind persons can learn to distinguish between differently coloured lights, not by the true differences of their respective colours, but by differences in their apparent illuminating power. The man colour-blind for red does not know the green from the red light because one is green and the other red, but because the retinal impression produced by the green is more vivid than that which is produced on him by the red. There may also be an acquaintance with some special effects or differences in quality acquired by repeated observation which may help the colour-blind, that persons who possess a naturally normal sense of colour are unable to appreciate. In this way colour-blind persons may pursue their callings for a certain time with impunity, notwithstanding their defects, which, however, may lead to dangerous results at any moment. When only a single coloured object, a flag or lamp, is exposed to view, no opportunity of contrast as regards illumination is afforded, such as a colour-blind person may turn to account when several objects of different colours are presented to him. Many atmospheric conditions and other circumstances may cause the clearness and intensity of light of particular coloured signals or lights to differ, or may make them appear to differ, in these respects, and as these variations in luminous intensity take the place of differences of colour with the colour-blind persons just referred to, it is obvious that no safe reliance can be placed on such substitutes for the true colour-sense. The importance of not placing reliance on any substitutes of the kind is forced on the attention all the more from reflecting on the very serious nature of the issues which must depend under a variety of circumstances upon the observers concerned in them possessing a full and correct appreciation of the colours exposed to view.

Holmgren's Test.—The test now generally adopted, or some modification of it, is the one introduced by Professor Holmgren, of Sweden. Holmgren's test is simple and easily applied. A number of skeins of wool of different colours are collected together. They consist of the colours corresponding with those of the spectrum, viz. red, orange, yellow, green, blue, indigo, and violet, together with yellowish-green, greenish-blue, purple, rose, brown, and grey. There are several shades of each colour, and at least five gradations of tint, or degrees of saturation, of each shade of colour, from the deepest to the lightest. The skeins of wool are mixed together in a heap on a table, exposed fully to daylight, not artificial light, and the examiner takes from it one of the skeins, and places it at some little distance from the general heap. The person under examination is then requested to pick out the other skeins which most resemble the test-skein in colour, and to place them side by side with it on the table. A person gifted with normal power of distinguishing colours will respond to the test easily and quickly; while another person whose perception of colour is faulty, or imperfect, will exhibit hesitation and will make mistakes in the colours, or shades of colour, of the skeins which he selects. A succession of similar tests may be applied in the same manner. Either very light or very deep shades of colour are recommended for the test-objects. The person under examination is not required to *name* the colours. His chromatic perception is judged by the manner in which he responds to the test. If faulty colour-sense be exhibited in the performance of the task, the tests must be continued in order to determine the kind and extent of colour-blindness, whether it be simply defective perception, or total absence of perception, of either red, green, or violet, or of any two of these colours.

To ascertain the Kind and Extent of Colour-blindness.—The following is the order of proceeding to establish the diagnosis as prescribed by Holmgren :—

1st Proof. A skein of pure light-green wool is given as the test. The specimen should neither be 'blue-green' nor 'yellow-green.' On examining the skeins which have been selected as similar in colour and shade, if they have been selected readily and found to be all of green tints, chromatic perception is 'not faulty ;' if after hesitation and difficulty in selection there is an approach to confusion of colour, the 'colour-sense is weak ;' if complete confusion of colour, if the green is matched with grey, brown, purple, or even with red, the conclusion is that there is 'colour-blindness.'

2nd Proof. A skein of purple colour, that is, a mixture of red and blue, is now given as the test. If all or the greater part of the skeins selected as being of the same tint are found to be so, the colour perception is 'not faulty ;' if only blue and violet wools have been selected and joined to the purple test-skein, in other words, if the person tested sees the purple only as blue, there is *colour-blindness for red* ; if green or grey, or both, have been selected as specimens of the same colour and shade as the purple, the person under examination is 'completely blind for green ;' if he join red and orange skeins to the purple wool, he is 'blind for blue and violet.' If wool of all colours, of the same shade or degree of saturation, should be taken as being alike in colour, there is evidence of complete and 'total colour-blindness.'

The proofs above given Holmgren regarded as sufficient for all ordinary cases, but for special individuals, as for official persons on whose right perceptions of colours important issues depend, he considered it advantageous to take further proof in confirmation of the conclusion arrived at from the foregoing tests, and then used a skein of pure red wool as the test-object. The colour-blind show by their selection that they confound green and red together, the 'red blind' persons particularly selecting, in addition to red, shades darker than the test-skein, such as shades of deep green, olive, and dark brown ; 'green blind' persons shades lighter than the red.

Thomson's Application of Holmgren's Test.—Dr. Thomson, of Philadelphia, has devised an instrument based on Holmgren's plan of testing colour-blindness which can be used by any intelligent observer for noting and recording whether a person is colour-blind or not; the ophthalmic surgeon being only referred to in cases which appear doubtful, or for confirming the observations when colour-blindness is stated to exist. This is reported to have been largely used in testing the colour vision of railway employés, with the result of discovering about 4 per cent. to be colour-blind. In this instrument 40 skeins of coloured wool are employed, the tints selected being those of the three test-colours, green, purple, and red, together with the usual 'colours of confusion.' The skeins have each a small metal plate attached bearing a particular number, which can be quickly read off and noted by the examiner. One half, or 20 skeins, are devoted to the green test, the 10 odd numbers being shades of green, the 10 even numbers 'confusion colours,' as greys, light browns, and others; the skeins from 21 to 30 have purple tints on the odd numbers, and blue tints on the even numbers; the skeins from 31 to 40 have red tints on the odd numbers, and brown, sage, or dark olive tints on the even numbers. The numbers are concealed from view of the person under examination, who merely selects the tints which seem to him to correspond to the test-tint given to him, on the same principles as have already been explained; while the examiner to whom the numbers are accessible simply notes on a proper form the numbers of the skeins, whether odd or even numbers, which have been selected. Reports thus obtained, and subsequently subjected to professional scrutiny in the cases of those who appeared to be colour-blind, elicited the proportion of 4 per cent. above mentioned. Holmgren concluded from his inquiries that the percentage of persons colour-blind was a little under 3 per cent.

Maréchal's Lantern Test.—A convenient method of examining for colour-blindness has been adopted in the naval schools of France on the suggestion of Dr. Maréchal, Principal Fleet Surgeon. It is a test by artificial light, and the manner of applying it is similar to that followed in the trials by Holmgren's wools. A lantern is arranged in such a way that a variety of colours may be shown by the same light on shifting the glasses. The examiner and the person under examination each have a lantern, and on the former showing a particular light the latter has to show a light of the same kind. Holmgren's wools are also used as tests by daylight, and the same mode of examination is carried out with pilots and signalmen employed in the French navy. Coloured signal lights, in addition to tests by daylight, are used in various countries in the examination of officers and men of the State navies. The same tests are applied in some instances to men belonging to the mercantile marine.

Plan for Testing Colour-sight in the Royal Navy.—All naval officers (medical officers only excepted) and sailors now enter the

Royal Navy as boys, and, I am informed, they are invariably tested as to their perception of colour, and that any notable deficiency in respect to it is regarded as a sufficient ground for rejection. The following is the order given in the 'Queen's Regulations and Admiralty Instructions' regarding the examination of candidates for admission into the Naval Service or into the Royal Marines as regards colour-sight. 'Whenever test-types are supplied, the power of vision of each eye separately, as well as together, is to be ascertained. If the persons under examination fail to distinguish the colours' (of the coloured test-types), 'they should be tried with brighter and decided colours; for this purpose red, blue, green, and yellow ribbon flags may be used.' Coloured wools are in use at the Admiralty for doubtful cases among officers.

Influence of Colour-blindness in the British Military Services.— This affection, which is very important as regards persons who have to depend on coloured signs for guidance, such as signal officers, look-out men at sea, and railway officials, or as chemists and others concerned in practical analysis, in which the presence of particular ingredients is determined by such tests as coloured solutions, or by the effects produced so far as colour is concerned; as regards physicians in the diagnosis of certain diseases character- ised by colour, as *scarlatina*; and even as regards officers in general command who require to be able to recognise clearly and quickly the colours of uniforms; [1] was not officially defined to be a disability in candidates for commissions as officers in the British army until the regulations promulgated on September 1, 1887, were issued. It is laid down in these rules that ' inability to dis- tinguish the principal colours ' will cause the rejection of a candi- date. It is not ruled to be a disability as regards soldiers in the ranks, nor are there any orders for the colour sense of men brought as recruits to be tested by examination. Neither has it been in- cluded in the lists of disorders of the eyes incapacitating men for mili- tary service either in our own or, I believe, in any foreign armies. It is important for medical officers not only to be acquainted with the nature of colour-blindness, but also themselves to have

---

[1] I was once aware of an officer discharging the duties of a musketry in- structor who could not distinguish the red flag or ' danger signal ' at target practice by its colour, but only by its larger relative size ; and of a general officer who was necessitated from the same defect to ask a member of his staff whether a body of troops were dressed in red or blue. I have also known a surgeon in civil practice who was temporarily affected with achromotopsia, during which period he could not distinguish red colours. A patient affected with scarlatina appeared to him to be of a yellowish tinge, as if he were suffering from jaundice. The late Professor de Chaumont informed me that difficulties had arisen in some instances among the surgeons on probation in the execution of practical analysis in the Hygienic Laboratory of the Army Medical School from the fact of the surgeons concerned being colour-blind ; and Professor Sir Wm. Aitken has met with similar cases in the Pathological Department on occasions when the nature of specimens seen under the microscope has had to be determined by the colour resulting from the applica- tion of certain reagents.

perfect colour-sense, and to be familiar with the methods of diagnosing colour-blindness in its several varieties, for they are the persons whose duty it will be to decide whether candidates for commissions are able to distinguish the principal colours accurately, and whether defective colour-sense exists or not in any special cases that may be submitted to their judgment.

The serious nature of the defect is acknowledged as regards officers and sailors of the Royal Navy, who have to distinguish accurately signals made by flags and pennants, for the power to execute this duty aright depends more upon correct appreciation of the colours than of the forms or even the markings of the objects presented to view. The recognition of the colours has to take place quickly and readily, too, under conditions in which their brightness is more or less impaired by mist and other atmospheric impediments. A colour-blind officer or sailor could not recognise the distinguishing pennants of ships, nor interpret orders conveyed by the usual combinations of signalling flags, for many of them have the same shapes and dimensions and only differ in their colours. This remark equally applies to most of the flags in use for signalling, as well as to the coloured lights, red, green, or white, which are shown by vessels at night to indicate their position and the direction which other vessels are to take in steering past them. The ill-consequences of colour-blindness in such cases may not be confined to mere failure of recognition of the objects concerned, but in many instances an erroneous interpretation of them induced by the defect may lead to most grave and irremediable calamities. It is not to be forgotten, moreover, that congenital colour-blindness is an incurable disorder, and that while the use of glasses and such optical instruments as the telescope may make up for some visual defects, they cannot lessen or in any degree correct the defect of achromotopsia. Candidates, therefore, for service in the Royal Navy, both officers and men, require to be carefully examined in respect to their faculty of distinguishing colours. The colours employed for the pennants and flags used during daytime in the Royal Navy of Great Britain are the three colours, red blue, and yellow, together with white and black, while at night, red, green, and white lights are used, so that it is for these colours that candidates chiefly require to be tested. At the same time it is not to be forgotten that, in addition to the simple colours named, compound colours are used in the flags of foreign navies.

Indian Government Services.—Candidates for admission into the public services under the Government of India are disqualified for service in various departments by the existence of achromotopsia or dyschromotopsia. The particular branches of the Indian service in which the regulations describe colour-blindness to be a disqualifying defect are named in Chapter IX.

Amblyopic Colour-blindness.—This is characterised in some cases by imperfect perception, or loss of perception, of certain colours, the results of morbid conditions of the retina and optic nerve. Abnormal sensations of colour, such as the prevalence visually of a tint of some particular colour independent of whatever may be the actual colours of the objects presented to view, should be distinguished as 'coloured vision' rather than as 'colour-blindness.' Xanthopsy, or the yellow vision, produced by the administration of santonin, is an example of the visual condition referred to. Red, in amblyopic, as in congenital achromotopsia, is the colour which is most frequently not discerned, while the power of distinguishing yellow or blue remains the most persistent. Amblyopia with central scotoma seems to be particularly liable to be attended with impairment of perception of colour. Optic neuritis, neuro-retinitis, and any diseases inducing atrophic changes in the optic nerve have the same tendency, and an alteration in the right perception of colours sometimes appears very early in their course. The disabling influence of this form of colour-blindness as regards military service depends on the amblyopia and morbid states with which it is associated rather than on the colour-blindness or colour-perversion itself.

CHAPTER VII.

ACUTENESS OF VISION.

THE term 'Acuteness of Vision' or V. has reference to quality of retinal sensibility, and might strictly be limited to the functional capacity appertaining to the impressible retinal elements; but as this capacity can only be practically tried or indicated by a measure of that on which the retinal function is exercised, the term as usually employed is applied to the size of the retinal image which can be visually recognised, or, what amounts to the same thing, the size of the object, which, under favourable conditions, can be seen and distinguished by the eye at a certain determinate distance.

Normal Acuteness of V.—The term 'normal acuteness of vision' is generally understood to imply clear recognition, in broad day-

light, of an object toward which the eye is directed under the
smallest visual angle for which the eye in a state of health is
organised.   This can only be regarded as an approximate definition
of the term, for the illumination of objects varies as the intensity
of the daylight to which they are exposed, and this differs greatly
in different countries and under different states of weather.   As
an optical expression, normal or full acuteness of vision refers to
that degree of perfection of which we have experience in the
human eye; it seems beyond doubt that some animals, such as
certain birds, can see objects of their own size from distances beyond
those at which they themselves can be seen by man, and it is quite
conceivable that a higher degree of visual acuteness than any
hitherto observed might be attained in the human eye through
further development of its visual organisation.   The acuteness of
vision referred to is central, or that which exists at the region
of the macula lutea.   Complete acuteness of V. can only exist
when the organic textures and transparency of the media of the
eye are perfect; when its refractive power is accurately adapted to
the distance at which the object looked at is placed; and when the
retinal sensitiveness, or the functional energy of the retinal elements, and the nerve conduction are also perfect.   An eye may be
emmetropic, yet its visual acuteness defective at all distances owing
to cloudiness of some of the media, to a defective or disordered
state of some of the retinal elements, or to other morbid conditions; on the other hand, an eye may be ametropic, and possess
perfect V. at some distances though not at others, or at all distances
when the ametropia is corrected.

From the foregoing it is evident that V. may vary in degree
from normal acuteness to mere recognition of light of a certain
amount—quantitative vision, and, further, that the particular degree
of V. in any given case will admit of measurement if a definite
objective standard of normal V. be agreed upon.

Alertness of V.—The speed with which particular objects are
distinguished varies in different individuals whose degree of V.
may be found to be similar by ordinary observation.   Such relatively slow or quick sight may be best defined by the expression of
'alertness of vision.'   Increased alertness of vision accompanies
increased intensity of visual impression, as in binocular compared
with monocular vision, but appears also to depend on individual
habits and constitutional peculiarities, visual practice, and, perhaps,
also on relative perfection in regard to transparency of the ocular
media, and, consequently, relative speed in the formation of the retinal
images of objects.   As senile changes take place in the refractive
media and ocular tunics, alertness of V. diminishes, and to this
probably in some degree are due the cautious and relatively slow
movements of the old.   Measure of alertness of colour vision
becomes important under circumstances in which the time for

coloured signals to be acted upon is very short, so that dyschromotopsia, though not a completely disabling defect like achromotopsia, is still one of serious account under some circumstances. Under certain conditions, as in narrow channels, the green and red lights of ships approaching each other rapidly from opposite directions may not be visible until the vessels are at a relatively short distance from each other, and a single minute, or even less, may be all the time available for the look-out man in each ship to take note of the lights and give the necessary steerage for passing so as to prevent collision.

Relatively perfect alertness of V. is of the utmost value when rapidly moving objects, coming suddenly into view, have to be recognised and aimed at on the instant. The vital importance of visual alertness, of rapidity, as well as of precision of aim, has been forcibly demonstrated in recent wars, especially in some of those in which British troops have been engaged against agile and unfettered foes such as the Zulus and other native tribes in South Africa, and as the Arabs in the Soudan. Alertness of V. is a quality which in emmetropic and otherwise normal eyes may be largely developed by well-directed training and practice.

Relative Acuteness of V.—Relative acuteness of V. is conveniently determined by the relative sizes of the visual angles under which objects of known magnitude can be perceived by different eyes. The power of separately discriminating visual impressions varies with the measures of the visual angles under which a series of objects, and the intervals between them, can be clearly seen; In other words, the limits of distance at which objects, whether complex or simple, can be seen separately and at the same time sharply, one from another, vary according to differences in degrees of acuteness of vision. When a standard of average normal V. is fixed, it becomes easy to express the relative V. in any given instance by comparing it with that of the standard.

Measurement of Acuteness of V.—Instruments designed for measuring acuteness of vision are called 'optometers.' Optometers can also be employed as means of measuring the range of accommodatory power. They consist of various objects (optometric test-objects) by which the amount of visual power possessed by an eye in its natural state may be tested, and of means associated with them, of measuring the distances at which these objects, when they can be clearly discerned, form defined images on the retina.

They can be readily turned to account not only for ascertaining permanent states of visual acuteness, but also for observing changes in degrees of visual acuteness at various intervals of time. The measurements are either taken in inches and parts of inches, or in metres and parts of metres.

Central Acuteness of V.—Although perception of the prevailing light is diffused over the whole field of vision, V., or recognition of objects, is sharpest, in the normal state of the eye, at the posterior pole of the visual line in the region of the macula lutea, and lessens in degree from this central part of the retina in proportion as its periphery is approached. In using an optometric test the eye is fixed upon it in the same way as it is naturally in observing any other object. It is, therefore, the *central acuteness of vision* which is ordinarily measured by optometers. Although ex-centric loss of V. rarely exists without a certain amount of loss of centric visual power, still, in some amblyopic states it is necessary to ascertain particularly the V. of ex-centric parts of the retina; the observations have then to be taken by other means, viz. by the use of perimeters for determining the form and limits of the field of vision, and the degree of visual perception over its area.

Optometric Test-Objects.—The test-objects employed in the measurement of V. are of various kinds. Among those in ordinary use are fine hair-wires, and objects of generally known forms, such as printed letters and numbers. Optometric objects are used for testing the distinctness of both near and distant vision. In testing the near point of distinct vision the objects employed for optometric purposes should, as a general rule, be of very small dimensions. The retinal image of a small object, from its extreme minuteness, ceases to be recognisable if it be blurred by diffused rays, while a larger image, notwithstanding a certain absence of definition from circles of diffusion, may be recognised from its mere size. In testing the far point of distinct vision it matters not what the size of the object is provided its size has a definite relation to the distance at which it is placed, and that the visual angle which it subtends with the eye is known to the person conducting the test.

Hair-wire Optometer.—This instrument consists of a series of very fine black wires, fixed in a frame which can be moved along a groove on a scale-board between 30 and 40 inches long. There are two separate grooves on the board and two wire optometers for the two eyes. At one end of the board are two openings through which the eyes regard the wire frames; they consist of grooved wire receptacles into which lenses may be placed at pleasure. The measurement scales on the board are in inches and parts of inches, and in centimetres. The wires can be shifted in position within the frames so that either a vertical, horizontal, or any intermediate direction can be given to them. In testing for the near point of distinct vision the limit of nearness is found at which the wires can be separately seen with perfect definition, and the distance is then read off on the scale-board. A like proceeding is adopted for testing the far point of distinct vision. By changing the position of the wire optometer it may be ascertained whether the near or far points of distinct vision respectively are the same both when the wires are placed vertically and horizontally, or have any other two opposite directions given to them. If they are not alike, there is astigmatism, and the far and near points in the different positions being noted, the degree of astigmatism may be approximately ascertained.

In practice it is found that the hair-wire optometer answers well for intelligent and observant persons, but frequently gives rise to difficulties when uneducated persons are subjected to trial by it. It is by no means easy to get some persons to decide when perfect definition of the wires is obtained and when the definition first becomes imperfect. Letters that we know must form defined images when they are easily read, and of which we know the images are not defined when they cannot be read, or are imperfectly defined when they can only be read with difficulty, constitute more simple and reliable tests for such persons.

The printed test-types in common use are of two kinds, Jäger's test-types and Snellen's test-types.

Jäger's Test-types.—These consist of paragraphs printed in the differently sized types which are in ordinary use for printing in different countries. The sentences selected are numbered according to the sizes of their types, the sentence in smallest type being distinguished as No. 1. The numbers increase as the sizes increase. They were at first arranged with a view to overcome the difficulty experienced by ophthalmic surgeons of different countries in understanding the nature of the letters referred to by one another as objects which patients were able to read under special circumstances, whether under different refractive states, or after operative proceedings or other treatment, owing to letters being only then distinguishable by technical names arbitrarily adopted among printers and differing in different countries. To get rid of this difficulty Professor Jäger, of Vienna, arranged the types used in different languages according to their corresponding sizes, and distinguished them by numbers as above mentioned. Jäger's test-types in consequence received the names of *types of universal reference.*

The individual letters are not framed on any common principle. Although there is a general correspondence of size in the type according to the number assigned to it, particular letters differ in their dimensions from each other. Some letters differ from others in width or height, and some strokes differ in thickness, whatever the number of the type. Various other well-known differences exist among the letters. The application of these types as accurate tests of V. is in consequence necessarily imperfect and limited. They are, however, occasionally useful from their forms being familiar to readers, so that their recognition is not interfered with as far as any peculiarity of shape is concerned. But such letters would manifestly be more generally serviceable if they were fashioned on such a fixed basis that every one of them when placed at a given distance presented the same visual angle, and had definite relations in respect to size with all the other types. Snellen's types were designed for the purpose of meeting these requirements.

Snellen's Test-types.—As Snellen's test-types form the standard by which visual acuteness is tested by medical officers of the British and Indian public services, and as they are the tests generally employed by ophthalmic surgeons in civil practice, their nature and peculiarities ought to be well understood.

Unlike Jäger's test-types, they consist of specially formed letters, all fashioned on one and the same fixed basis. The sets of types are of different sizes, and each set is accompanied by a special number. These numbers bear definite proportions to each other, and to the sizes of the letters with which they are connected.

When Dr. Snellen arranged these types he experimentally determined the smallest visual angle under which letters could be read provided the vision of the reader were of normal acuteness, or, in other words, the least magnitude of the retinal image of a letter which enables that letter to be distinctly perceived, for on the size of the visual angle, as elsewhere explained, depends the size of the retinal image.

Dr. Snellen has taken as his standard that an emmetropic eye of normal visual acuteness can perceive a plain rectangular object in fair daylight when it subtends a visual angle of the 60th of a degree, or one minute. This is the smallest object that can be seen—the *minimum visibile*—according to Snellen's standard. The space such an object would occupy in a circle of 24 inches in diameter, of which the eye may be supposed to be the centre, would be about the 285th part of an inch of its circumference. At a distance of 12 inches therefore, on this basis, the eye can perceive the presence of a plain object about 285th of an inch in size.

But though a uniform object can be seen by the eye when occupying only a space of the 60th of a degree, a complex object, though visible, cannot be recognised under so small a visual angle. The smallest visual angle permitting clear recognition of such broken and irregularly formed objects as printed letters, according to Dr. Snellen's standard, is one-twelfth of a degree, or five minutes. Thus at a distance of 12 inches, the eye can recognise a letter about the 57th of an inch in size, or, according to Snellen, a letter 0·209''', or 0·0174'', $\frac{1}{57\cdot47}$th of an inch, in dimensions, Paris measure.[1]

[1] Dr Snellen has taken 0·209''', French measure, or 0 0174'', for the dimensions of each of his letters and of the spaces between them, which are to be recognised at a distance of 12'' by a normal eye. At the distance named, viz. 12'' from the cornea, or 12·28'' from the nodal point of the eye, such an object subtends a visual angle of about 5 minutes. The dimensions of the retinal image of an object of this size would be 0·00089'', or $\frac{1}{1123}$; but if for a letter having a visual angle of 5 minutes a plain object, such as a test-dot, with a visual angle of only 1

It is on these principles that Dr. Snellen has arranged his test-types. They consist of separate letters, quadratic in shape, or occupying a space the linear boundaries of which form a square, and are all formed of strokes, or *limbs* as they are called, the breadth of each of which respectively is one-fifth of the linear dimensions of the square within which the whole letter is contained. The types numbered on the duodecimal system bear numbers from I to CC according to their sizes ; No. I being the smallest, No. CC the largest among them. These numbers also express the number of feet at which the types can be read by an eye possessing normal acuteness of vision. When the letters are read at the fixed distances in feet, which are numbered above them, the eye, in seeing a limb of the letter, is seeing an object which occupies an angle of one minute, while in seeing the whole letter it is seeing an object which occupies an angle of five minutes in the visual field.

Dr. Snellen has supplied sentences in various languages printed on the same principles as the separate letters, and like them bearing numbers in accordance with their sizes, but it must not be forgotten that when printed sentences are used as tests, some words may be read although the separate letters are not seen distinctly, from the fact of the eyes having become familiar with their general aspect, owing to their frequent occurrence in printed books. Wrong inferences might be drawn from this fact. Separate letters of definite size are not open to this objection ; all the parts or limbs of each letter or figure must be imaged with a fair average of distinctness, or the object could not be recognised. Moreover, the intervals between the separate letters, numbers, and test-objects can be arranged to be equal in dimensions with those of the objects themselves, whereas no such systematic plan can be carried out when the letters are grouped in words and sentences.

Arithmetical numbers of various sizes are included among the test-types for persons who cannot read letters but can decipher such figures ; and for those who cannot distinguish either letters or numbers, simple objects such as circles, lines, crosses, squares, and others are added. They are drawn to scale on the same principles as the letters, and are intended to be used in the same way.

It will be observed from the foregoing description that Snellen's types have not been designed for giving *qualitative* estimates of V., although they may be partly applied, as elsewhere explained, to such a purpose, but have been principally arranged as means of obtaining *quantitative* estimates of visual power.

minute (0·00348″) be substituted, the dimensions of the retinal image would be 0·000178″, or $\frac{1}{5614}$″, and this would approach the size which physiologists have assigned to one of the retinal elements, or cones, at the site of the macula lutea. As many persons are able to see objects under smaller visual angles than that adopted by Dr. Snellen, it is evident that the approximation to the size of one of the retinal elements referred to, variously given from about $\frac{1}{7000}$″ to $\frac{1}{9000}$″, would be more closely attained.

Snellen's Types on the Metrical Scale.—The test-types for the determination of the acuteness of vision which have been, and still are, in general use in the British service were officially distributed to the army medical officers in the year 1864, shortly after that edition was published. In an edition published in 1882, Dr. Snellen adopted the metre as the standard of unity, and since that date all his test-types, or optotypes, as they are called by him, and other optometric objects, have been numbered on the metrical system. They bear figures above them which indicate the numbers of metres, or parts of metres, at which the test-types should be read by an eye possessing average normal acuteness of vision, in the same way as .the former test types were marked in Paris feet and inches. There is no change in the formula for expressing the acuteness of vision : it is still $V. = \frac{d}{D}$. The distances for which the types are arranged vary from the smallest, in which D is equal to 0·5 m., to the largest, in which D is equal to 60 m. The smallest type should therefore be read at a distance of half a metre or about 20 English inches, the largest at 60 metres or about 200 feet, by an eye endowed with normal acuteness of vision. In these later editions some improvements have been made in the tables for detecting astigmatism, and in the tests for acuteness of colour-sight. The edition of 1864, in which the types are numbered in feet and inches, still remains the edition in general use in the British service, but the edition of 1885 has been recently ordered to be used for testing V., in the examination of commissioned officers of the army.

Burchardt's International Sight-tests.—These test-objects have been formed on principles which differ in some respects from those of the preceding optometric objects. They are intended to enable surgeons to ascertain the near and distant points of distinct vision in persons submitted to their observation, as well as the existence of astigmatism if it be present, without the aid of lenses. They were first published in 1870, but a larger edition (third) was published in 1883.[1] The purpose of Dr. Burchardt was to get rid of the objections to the use of letters—viz. that they are only applicable to men who know how to read, that some letters are easier to be recognised than others, that this recognition involves mental effort as well as the act of seeing, and that the upright strokes of the taller letters are often not distinguishable at the same distances as the cross strokes of the shorter letters, owing to astigmatism. The test-objects for acuteness of vision employed by Dr. Burchardt are black discs of different sizes on a white ground. I had already called attention to the use of such discs for military

---

[1] *Internationale Sehproben zur Bestimmung der Sehschärfe und Sehweite.* Herausgegeben von Dr. M. Burchardt, Oberstabsarzt 1. Kl. &c. Dritte verbesserte und vermehrte Auflage. Kassel, 1883. (International Sight-tests for determining Acuteness and Range of Vision. By Oberstabsarzt 1 Cl. Dr. M. Burchardt. Third edition, improved and enlarged. Cassel, 1883.)

purposes in the Army Medical Reports for the year 1860.[1] My arrangement of the sets of discs was such that they corresponded in their diameters with the series of Jäger's test-types; but they were very badly printed, and were needlessly scattered by the printers, in a manner never intended, over the pages of the paper which they were designed to illustrate. Dr. Burchardt has greatly improved on the discs just referred to, and has grouped them together on principles of his own. Instead of measuring the acuteness of sight by the visual angle under which the separate objects are recognised, Dr. Burchardt measures it chiefly by the visual angles of the intervals between the objects. Just as the sensibility of the skin, or sense of touch, may be measured by the limits of distance at which the two points of a pair of compasses are separately felt, so, on Dr. Burchardt's system, the sensibility of the retina, or visual acuteness, is measured by measuring the distance between two objects, or between their retinal images, necessary for their separate perception. It is therefore the minimum interval of separation of objects appreciable by the retina that constitutes the test of acuteness of vision in Dr. Burchardt's international sight-tests. Thus, for example, Dr. Burchardt shows that discs of 0·1 mm. diameter, placed in a row of intervals of 0·1 mm. from each other, and viewed from a distance of 60 cm. (24″), are perceived as a continuous line; at a distance of 20 cm. (8″) appear as a rough line with occasional swellings; at 16 cm., or a little more than 6″, are recognised separately and can be counted. That the power of the eye to count objects of simple forms depends not alone on the visual angle which the objects severally subtend, but partly on the lengths of the intervals between them, has been shown in a very simple manner with the army test-dots ⅛″ square by Inspector-General Dr. Lawson. When the intervals between them were each of the same size as a test-dot, viz. ⅛″ square, the dots could be counted by himself at a distance of 36 feet; when there were two such squares, at 58 feet; when three squares, at 74 feet; and the same dots with intervals of four such squares could be counted at 82 feet distance. These experiments prove that, within certain limits, the distances at which the test-dots can be clearly distinguished will vary with the distances at which they are placed apart.

All the sight-tests in Dr. Burchardt's tables have diameters which are 1,600 times less in length than the length of the distance at which they are to be seen by an eye of normal acuteness of vision according to his standard. Thus the discs in the 60 metre table have each a diameter of 37·5 mm.; those in the 1,600 mm. table have a diameter of 1·0 mm.; while the discs of the 10 centimetre

---

[1] Notes on the Examination of the Visual Fitness of Recruits for Military Service, with special reference to Instruction in the Use of the Rifle. "Army Medical Reports," vol. 2 for 1860. London, 1862, p 462.

table have diameters each of $\frac{1}{16}$th of a millimetre. At the distances named each of the discs, and each interspace between every two adjoining discs, subtends a visual angle with the eye of 2·15 minutes. This is 1·15 minutes larger than the visual angle under which the test-dots on Snellen's standard are required to be counted by an eye reputed to possess normal acuteness of vision.

Dr. Burchardt in the last edition of his sight-tests (1883) has adopted the metric system of measurement. The figure of distance attached to his largest discs is 60 metres, and the figures descend to 10 cm., the sizes of the discs decreasing in proportion. He has also added two sheets of block-letters, graduated in size, for distances from 20 metres to half a metre, as well as a sheet of large discs designed for determining at a distance the directions of the faulty meridians in cases of astigmatism. Some of the tables are reduced by photography from accurate drawings, and are clearly engraved on card tablets of pocketbook size, so that they are very portable, while the back of each card bears concise instructions on the manner of using them in the detection of true as well as simulated differences of visual acuteness and refractive power. Special small cards and tables for the diagnosis of astigmatism are added. They are thus conveniently arranged for fulfilling their purpose; but, on the whole, Srellen's test-types, although in some respects less scientifically accurate than Burchardt's sight-tests, appear to be more generally serviceable, and, as they have been sanctioned for employment by medical officers of the British and Indian services, while a knowledge of reading is yearly increasing among all classes of men in England, there appears to be no sufficient reason for introducing others into use.

Snellen's Standard of Visual Acuteness.—In practice it will be found that particular eyes, especially sound young eyes, have a considerably higher degree of V. than the standard fixed upon by Dr. Snellen. An object which subtends an angle of only half a minute, or even one-fifth part of a minute or 12 seconds, when directly illuminated by the sun, is visible to some eyes. Colour has an influence : a white object with the light of the sun shining upon it may be seen under an angle of 12 seconds, but, under the same circumstances, a similar object, if red in colour, would only be seen under an angle about double that size. Sudden change in the intensity of the light to which the eye is subjected, ocular fatigue from prolonged visual efforts, pressure on the globe, and a variety of other causes, will temporarily interfere with the power of reading the types at their normal distances, and, unless taken into account, may lead to an erroneous conclusion in a given case that the visual acuteness is below Snellen's standard, when it really is not so. Some persons can read test-types at distances considerably beyond those indicated by their accompanying numbers—when, therefore, the angle under which each type is recognised is less than an angle of five minutes. I have seen them read, under favourable conditions, at double the indicated distances—when, therefore, the visual angle has been reduced to about $\frac{1}{24}$ of a degree, or $2\frac{1}{2}'$, and V. has been = 2, or twice Dr. Snellen's standard. Their recognition at a distance of even thrice Snellen's standard in a good light has been recorded. Snellen's test-types, as numbered, are consequently to be regarded not as absolute standards of perfect V., but rather of *average* normal acuteness of vision, as deduced from actual observation of a large number of persons free from visual defects; while those in whom V. is found to be twice or thrice Snellen's standard are to be regarded as exceptions to the general average of normal visual acuteness.

Uses of Snellen's Test-types in Military practice.—The great value of these test-types is the ease and readiness with which they can be used for practical purposes in ocular examinations. It is not of so much moment that the standard on which they are based shall be accurate in its estimate of normal V. as that the types can be used for determining whether the acuteness of V. in any given case is equal to, below, or above that standard. At the same time, Snellen's standard may be accepted as a fair standard of normal V. under the ordinary conditions of everyday life in Europe.

As the letters are all formed on one and the same principle, they are capable of being applied to various uses in examinations of visual acuteness. Being all seen under the same visual angle at the distances indicated, they all at those distances have the same apparent magnitude; and, as they are all formed in the same fashion, and occupy proportionate areas, so also at the distances indicated they not only have the same linear dimensions, but also the same apparent superficial magnitudes. Letters of any one size may, therefore, in practice be substituted for letters of any other size within the limits of distance for which the eye is adjusted, or can adjust itself by the exercise of accommodation, of course provided illumination and other conditions be preserved alike.

Again, if two or more of the types be held at other than the named distances, whether more remote from or nearer to an observer, the visual angles under which they are severally seen will still be alike so long as the distances at which the different types are placed are relatively in accordance.

Snellen's test-types afford a simple and practically a sufficiently accurate mode of expressing the degree of V. in any given instance. Snellen's formula is the following. If V. (vision) be used to express the acuteness of vision; D the distance at which the type appears under an angle of five minutes, or the distance named with the type used; d the utmost distance at which the type can be read by the person under observation: then $V. = \frac{d}{D}$. In this arrangement D is a fixed quantity, d a variable one.

Examples : The 20-feet types are read at 20 feet, the 30-feet types at 30 feet; then V. $= \frac{20}{20}$ or $\frac{30}{30} = 1$, or, in other words, is normal. If the 20-feet type can only be read when the person in approaching the types reaches a distance of 10 feet, the 30-feet type one of 15 feet, then V. $= \frac{10}{20}$, or $\frac{15}{30}$, or $\frac{1}{2}$, and is only one-half of the normal standard. Practically, in determining relative degrees of V. by these means, it is better to use one common test-type, the 20′ type, for example, as the standard.

A convenient mode of using Snellen's types is to have the scale of types, one row above the other, placed on a suitable stand in a good light at a fixed distance—a distance, for example, of 15 feet from the person under examination. In the formula V. $= \frac{d}{D}$, d then becomes a fixed quantity, and D a variable one. The person whose vision is under trial is desired to read the smallest row of types which he can see clearly at that distance. If he can read the 15-feet type, but none smaller, then V. $= \frac{15}{15} = 1$. If he can read the 12-feet type, then V. $= \frac{15}{12} = 1\frac{1}{4}$, or his visual acuteness is one-quarter above Snellen's standard. If he can only at the distance named read the 20-feet type, then V. $= \frac{15}{20} = \frac{3}{4}$, and his acuteness of vision is only three-quarters of Snellen's standard. The advantage of using a distance such as one of 15 feet is that there is no need, under ordinary circumstances, for exercise of accommodatory exertion on the part of the person under examination, for the rays of light from objects at this distance reach the eye practically as parallel rays, and it is only in the case of hypermetropes that accommodatory exertion will be employed.

The use of Snellen's types saves time in examining the quality of eyesight in any unknown case. If the person under examination reads with each eye the 20-feet type at 20 feet with ease, there is no ocular defect of sufficient importance to require further investigation. If it, and other types, can only be read short of their normal distances, some defect exists, and the necessity is indicated for further examination by trial lenses, or by the ophthalmoscope, in order to ascertain the cause of the deficiency of visual acuteness.

Either ametropia or lessened accommodatory power is indicated when some of the types are seen clearly at the distances marked above them, while others are not seen clearly at their distances. In such cases the refractive power of the eye does not maintain correspondence with the relations which are preserved between distance and size in the types. If an eye can read the 1-foot type at the distance of 1 foot, the $1\frac{1}{2}$-foot type at a foot and a half, but cannot read the XX-feet type at 20 feet, and other larger types at their respective distances, myopia is indicated; while, if the XX-feet type can be read at 20 feet, but the smaller types cannot be read at their distances, either presbyopia or hypermetropia is probably present.

If deception is attempted, whether of a positive or negative kind

—and the manner in which the replies are given will generally point clearly enough to the fact when such an attempt is made—it may often be exposed by subjecting the person under examination to tests by different but adjacent types. If there be no attempt at deception, but the alleged deficiency of V. be real, the relations between D and d will be preserved when types of different sizes, such as the 20, 30, or 40 feet type, are presented to be read. If V. = $\frac{40}{40}$, it ought to be equally $\frac{30}{30}$ and $\frac{20}{20}$ if other like conditions be carefully preserved; if different values be given to V., deception of some kind may be suspected. The smaller-sized types, Nos. 1 to 3 or 4, should be excluded from the comparison.

The degree of acuteness, equally with the alertness of V., are naturally lessened in the latter years of life, owing to decreased transparency of the dioptric media, decreased retinal sensitiveness, and other senile changes. But, from a table published by Inspector-General Dr. Lawson, comprising a series of observations on 974 persons, it would appear that V. gradually declines from a very early age, even as early as fifteen years. The decrease in the instances quoted was independent of diminished accommodation. It follows that other causes beside senile changes must be sought for to explain the lessening of V. with age, should Dr. Lawson's facts be substantiated by more extended observations.

According to a table quoted by Dr. Snellen, deduced from a series of observations by T. v. de Haan, of Utrecht, on 281 persons of ages varying from 7 to 83 years, the state of whose eyes had been previously ascertained to be sound and healthy, V. was above Snellen's standard up to the age of 40 years, but slightly declined between 20 years and the age named, became $\frac{1}{10}$th below Snellen's standard at 50 years, and was reduced nearly to $\frac{1}{4}$ at 80 years of age. The actual figures resulting from the experiments, according to the quotation, showed that the average of V. for ages up to 20 years, the figure 20 being used as a standard of comparison, was as 22·5 : 20; at 30 years was as 22 : 20; at 40 as 20·5 : 20; at 50 as 18 : 20; at 60 as 14·5 : 20; at 70 as 13 : 20; and at 80 as 11 : 20.

Snellen's test-types can be readily turned to account in the application of any rule that may be laid down as to a required standard of V. Thus, for example, a military friend gives me, as a rule, from the result of his experience, that a soldier, to be effective, must be able to distinguish clearly a man from any other object at a distance of 500 yards under ordinary illumination, that is, in a moderately clear daylight, and with no more striking contrast of background than what is met with in ordinary fields or moorland. A sentry on an advanced post who could not distinguish an enemy at that distance in front of him would endanger the safety of a force. With such a background as the 'skyline,' or any background forming a marked contrast with the object, a man ought to be recognised at 1,000 yards. The amount of light reflected from the object looked at, relatively to the amount of light reflected from the objects by which it is surrounded, and the character of the background, are always important elements in regard to visual perception, in addition to the size of the visual angle subtended by the object.

The rule for recognition at 500 yards may be applied by means of Snellen's types thus :—Assuming the height of a man to be that at which the height for infantry is calculated in rifle practice, viz. 6 feet, the visual angle under which he would be seen at a distance of 500 yards is 13' 44'', or nearly 2·7 times the visual angle under which Snellen's types are seen. Recognition of 20' Snellen on toned paper in an ordinarily lighted room at a distance of 7' 5'' may therefore be used as a test that one man is capable of distinguishing another man at a distance of 500 yards under the above-named conditions. A man 6 feet in height, to be seen under the same visual angle as Snellen's types, would have to stand at a distance of about 1,375 yards off. But practically at such a distance, owing to the effect of the intervening atmosphere and other circumstances, the man could not be distinguished, although an object having the same visual angle might be seen plainly in a nearer position under adequate illumination.

Measurement of Visual Acuteness when associated with Ametropia.—When a low degree of V. is due to simple uncomplicated ametropia, whatever the nature of the latter, if it be corrected by suitable lenses, normal V. will be restored ; if, however, amblyopia or other complications exist which participate in causing the degradation of V., the defective V. will be only partially improved by the lenses. Whenever, therefore, V. is found to be below the normal standard on first examination, and ametropia to be associated with it, the nature and degree of the latter should be ascertained with a view to its correction, and the eyes should be tested after the correcting lenses which have given the best possible results have been applied to them. In this way only can the true condition of the eyes, as regards their visual acuteness, be determined. The visual acuteness in all such cases should be noted before and after correction of the ametropic defect.

WEAK VISION.

There are three kinds of weak or defective V. which it is necessary for an observer to distinguish from one another in ocular examinations regarding visual acuteness. The first of these is amblyopia, derived from ἀμβλὺς, blunt, obtuse, and ὄψις, sight ; the second is asthenopia, from ἀσθενὴς, wanting in force, and ὄψις, sight ; the third may be designated nephelopia, from νεφέλη, a cloud, or mist, and ὄψις, sight—dimsightedness due to conditions by which the passage of light to the retina is obstructed or otherwise disturbed. The first is symptomatic of imperfection in the sensitive recipient or conducting nerve elements ; the second refers either to weakness in the internal structures which are engaged actively in adapting the dioptric apparatus to the varied requirements for clear vision at different distances, or to a deranged balance of power between them and certain external muscular motors of the organ ; the third is usually the result of morbid loss of transparency of some of the ocular structures.   It will be convenient to consider separately these three conditions, which are very distinct in their nature.

I. Amblyopia.

Definition.—Feebleness of vision from diminished nerve power. Impaired vision thus defined was formerly included with many other morbid conditions of different kinds under the general term ' amaurosis.' [1]   Amaurosis is now only used to express total loss of V. from annihilation of the function of the visual apparatus, generally due to intercranial disease, but due also to morbid changes of the optic nerve and retina.   Ambl., therefore, represents partial loss of visual sense, amaurosis complete loss of visual sense, as a result of morbid changes in the nervous apparatus concerned with the sense of sight.

Causes.—These may be either intrinsic, that is, due to diseased changes originating in the optic nerve itself, its cerebral connections, or retinal expansion ; or extrinsic, when the diseased conditions are induced in sequence to disease of neighbouring but functionally independent structures, such as cerebral tumours and other cerebral diseases giving rise to pressure on the optic tracts or involving them in the morbid processes, sequels of insolation, intraorbital tumours, diseases of the choroid and other intra-ocular affections, reflex irritation from branches of the fifth nerve, and others.   Amblyopia is also caused by a variety of constitutional disorders which lead to anæmia, impairment of nutrition, prolonged congestion, or to changes of the ocular nervous apparatus brought on probably by morbid or toxic materials circulating in the blood-vessels.   These include constitutional states of general debility due to repeated losses of blood from hæmorrhoidal or other sources, as also to excessive debilitating discharges, whatever their nature, to habitual and inordinate use of tobacco and alcohol, excessive

---

[1] Amblyopia is sometimes used to express the indistinctness of vision which is directly dependent upon the third variety named above, viz. diminished transparency in some of the anterior dioptric media.   It is practically more useful to limit its signification to diminished power of sight dependent on morbid conditions of the retina, optic nerve, or brain, without, however, laying down any limitations as to their nature or modes of origin, than to extend it to lowered acuteness of vision due to causes which are obvious to external observation, with or without the aid of the ophthalmoscope.

cinchonism, lead poisoning, secondary syphilis, diabetes, albuminuria, and a variety of cachectic conditions. Diminished retinal power from structural changes is one of the concomitants of old age. Other causes are mechanical injuries, such as blows about the orbit, producing optic paralysis, hæmorrhagic effusion, and retinal detachment. Sudden severe shock, or excessive intensity of light, as from a close flash of lightning, may be a cause of amblyopia or even complete amaurosis.

Lastly, just as retinal sensibility may apparently be increased in energy in a healthy subject by constant practice at natural objects, so it may be lessened by want of employment, *amblyopia ex anopsiâ*, as sometimes happens by mental suppression of the retinal image of one eye in strabismus. in some instances of anisometria, and also, when a corneal opacity exists in one eye, by the patient excluding this eye from binocular vision in order to prevent visual confusion. The retina, from these causes, as from any other which hinder an eye from taking its part in active visual exercise, gradually loses its susceptibility to impression, and its loss of sensitiveness becomes less capable of recovery in proportion to the duration of its abstention from functional employment. On the other hand, Ambl. may be induced by continued overstraining of the retina in prolonged work at minute objects, such as very small printed letters and figures, especially if the types and accessories are bad and the printing indistinct, as they are in some cheap modern reprints of standard works. The ill effects upon the retina are all the more marked, and occur the more speedily, when the person is placed under the influence of circumstances tending to impair his general health, and when at the same time his retina is overstimulated, and irritated by strong artificial and unsteady light, or by the biight glare of a tropical sun. Central vision under these conditions is at first weakened, while the peripheral portions of the retina retain their normal sensibility ; but, unless the necessary rest from the deleterious influences is taken, and suitable precautionary and remedial measures adopted, the peripheral, as well as the central, retinal sensibility becomes impaired, the visual field contracts, and, as occasional instances prove, the impairment may advance until all visual power disappears.

From the variety of causes, above enumerated, which lead to Ambl., it will be seen that it should be regarded rather as a symptom than as in itself a distinct disease. It is as a rule the negative to the positive expression ' acuteness of vision.' In many instances the diseases which give rise to Ambl. are unavoidably obscure, as when they are intra-cranial, so that the effect which is manifest, viz. the Ambl., can alone be distinguished and named. Again, in numerous cases where the loss of retinal acuteness is functional, as when it is due to anæmia, the excessive use of tobacco, and other similar causes, it frequently happens that no objective lesion can be observed under the closest examination. But in numerous other instances the cause of the Ambl. can be ophthalmoscopically demonstrated, and the special affection of which the Ambl. is a result should then be properly designated.

Symptoms and Diagnosis.—In its mildest forms the patient simply loses the power of perceiving very small objects clearly at any distance. But it may vary in degree from inability to see some of the smallest types up to inability to distinguish the types of largest size. The feebleness of vision may become aggravated until it is so weak that the patient is hardly able to see his way about, and in the end the sensitiveness to light may disappear and amaurosis be established. With a mild degree of Ambl. print of moderate size is held closer to the eye in reading than usual, in order to obtain larger images, and thus an inexperienced observer is liable to suppose erroneously that the patient is myopic.

Ambl. occurs independently of any refractive fault or diminution in accommodatory power. Either of these conditions may or may not be present concurrently with the Ambl. If a refractive defect be associated with it, whatever may be its nature, the correction of it will not lessen the existing Ambl., though in occasional cases the correction may tend, when associated with other means, to arrest its further progress. The effects of Ambl. are felt at all distances for which the eye is adapted, whether near or distant objects are regarded. There is not usually with Ambl. the sense of effort or fatigue that accompanies asthenopia, nor the 'blurred vision' of defective refractive power. If the eye be emmetropic naturally, or has been rendered so by suitable lenses, the amblyopic eye will still only be able to see objects under larger visual angles than are consistent with visual recognition by another which has normal acuteness of vision. There will often exist with it a contraction of the field of vision, and in some instances a weakened or deranged condition of the faculty of colour perception. Ambl. will, of course, be found to be accompanied by the characteristic symptoms of the particular diseased condition which gives rise to it, when the latter is of such a nature as to be definitely recognised. As soon as the fact of the existence of Ambl. is established, a true diagnosis of its cause should be sought for by a careful study of the history of the case, and by ophthalmoscopic investigation; for, if the exciting cause be one that admits of mitigation or removal, and the Ambl. be in an early stage, under appropriate remedies such as the nux vomica, and when such visual exercise only is permitted as can be carried on without effort, the restoration of normal retinal acuteness may in some cases be accomplished.

The following also are forms of Ambl. :—

(a) **Hemeralopia**, night-blindness, or that condition of weak vision in which the patient can see well in the full light of daytime. but cannot distinguish objects after sunset or in a dim light. This is frequently found among soldiers who have passed from a northern latitude to a tropical station. In these instances it is evidently due to temporary exhaustion of nervous power from over-stimulation by the bright light of the tropical day and the reflected glare from the water of the ocean, unrelieved by the variety of shade and colour which are met with on land. Hence the inability to perceive objects illuminated by the comparatively weak rays of moonlight. Snow-blindness appears to be of the same nature. The tendency to the occurrence of Hemeralopia among men who are exposed to its exciting causes, and the persistence of the impaired sensibility of the retina, will be greater in persons who have acquired a scorbutic taint, or who have become constitutionally debilitated from any cause. This description refers to simple functional hemeralopia; care must be taken not to confuse it with the diminished visual power which co-exists with retinitis pigmentosa, atrophy, and other structural changes of a grave nature in the retina. Perhaps, in addition to the glare, the exposure to the heat of the solar rays by day, and the alternation of chills from dews at night, may assist in the production of hemeralopia, for it has been observed to occur among many men together who have slept in the open air, as on deck, without overhead cover. Men who have suffered from hemeralopia appear to be specially susceptible to relapses on exposure to its exciting causes: hence the advantage of protecting the eyes of such persons by the use of coloured spectacles or goggles, to prevent its recurrence. Protection of the eyes from strong light is also a necessary part of the treatment of this affection among those who are actually suffering from it.

Hemeralopia is sometimes simulated, and is reported to be frequently assumed by soldiers in some foreign armies. Various stratagems have to be resorted to for the detection of the imposition if the condition be feigned, for when it really exists as a functional disorder there is no visible sign by which its presence can be proved. This fact should make medical officers very guarded in expressing an opinion that the disorder is simulated, however strong may be the suspicions which they are led to entertain on the subject.

(b) **Nyctalopia**, which is sometimes used as synonymous with night-blindness, really signifies the converse condition of hemeralopia, or that the patient can see better at night than he can during the daytime. In this state the disorder of the retina is shown by its being unable to bear the stimulus of bright light owing to hyperæsthesia. The normal acuteness of vision may not be materially lowered in subdued light; but strong light gives rise to trouble and confusion, induces various kinds of subjective luminous apparitions of which the patient finds it difficult to rid himself, while attempts to read print of moderate size or to examine objects in bright daylight produce all the symptoms of severe photophobia—ocular pain and dazzling, lachrymation, spasms of the eyelids, supra-orbital pain, and general distress. After sundown, or when the eyes are shaded by tinted glasses, the patient moves about with comparative comfort, and sees objects clearly that he could not distinguish in ordinary daylight. The intolerance to the bright light thrown on the retina by the ophthalmoscopic speculum sufficiently indicates the presence of this abnormal irritability; and this may happen in an eye where there have been previously no symptoms of inflammation, and in which the fundus seems to be quite free from inflammatory effects. Such cases are occasionally met with among the soldiers who are invalided for impaired vision from India. In these instances the affection seems to be due to the prolonged effects of tropical glare upon an over-sensitively organised retina, generally associated, however, with a lowered state of constitutional tone, and often with deficiency of choroidal pigment.

In the year 1885 a soldier of the Scots Guards was invalided from Egypt for symptoms of nyctalopia. He had had iridectomy performed in both eyes, prior to enlistment. He was probably passed as a recruit in consideration of his being in other respects a physically sound and finely proportioned man, and because at that time, in a subdued light in a closed room, he could count the test-dots at the required distance. He was useless in the open air in the daytime in Egypt. The irritating effects of the glare, owing to the loss in both eyes of the natural power of excluding excess of light by the action of the irides, caused a good deal of suffering at the time, and not only induced extreme difficulty of vision when the daylight was strong, which still existed on his arrival in England, but also led to a considerable amount of Ambl. On arrival at Netley, V. of the right eye was = $\frac{1}{5}$, of the left eye only = $\frac{1}{30}$ of Snellen's standard. The acuteness of vision was greatly improved in each eye in full daylight by the use of a diaphragm with a stenopœic aperture, but no means could bring it back to the normal standard.

(c) **Hemiopia, Half-vision.**—Impairment or loss of retinal perception, limited to the outer half of one eye and the inner half of the other eye (see fig. 73). The manner in which the fibres of the optic nerves decussate at the optic commissure suffices to explain how any cause which impairs the conductibility of either optic tract before it reaches the commissure, such as pressure from cerebral hæmorrhage, a tumour, or from any other source, may destroy visual power in the right or left halves of the two eyes, while the remaining portions of both retinæ retain their ordinary power of perception. Central vision sometimes remains unimpaired. The diagnosis can only be made out by noting carefully the field of vision of each eye. The restricted limits of the patient's view sufficiently show the existence of the hemiopia. If the left half of the field of vision of each eye is wanting, a lesion affecting the right optic tract and loss of visual power of the right half of each retina will be indicated, and *vice versâ*. Hemiopia, when confirmed, entirely unfits a soldier for duty in the ranks.

·   (*d*) **Scotoma.**—Partial deficiency or total loss of vision in an isolated portion or portions of the retina.   Scotomes are occasionally central, or are situated near the retinal centre, when serious amblyopia usually results ; or they may occur in ex-centric portions of the field, and vision be less interfered with.   The lesion ·may

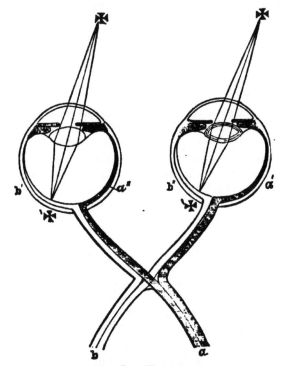

FIG. 73.—HEMIOPIA.

a, right optic tract, supplying *a'*, the right temporal, and *a''*, the left nasal halves of the retina.   Extent of lesion shown by the transverse lines. *b*, left optic tract, supplying *b'*, the left temporal, and *b''*, the right nasal halves of the retina, remaining unaffected, so that the field of view is here limited to the left half of each retina, as indicated by the objects ✠ ✠, and their images at ✠' ✠'.

occur in one eye singly, or both eyes may be affected simultaneously. The defect may be a sequela of some local lesion, of retinitis, optic neuritis, or of some form of choroiditis. A dark spot or spots, often surrounded by a margin of haziness, appear in ordinary vision, and correspond with the portions of the retina that have lost sensibility to light. The spots move in concert with the movements of the eyes. Besides the amblyopia, central scotoma is usually accompanied with localised disturbance of the power of distinguishing colours, the retina around retaining colour perception in integrity. In some cases a part of the retina may be atrophied, separated from the choroid, or may be seen on ophthalmoscopic examination to be infiltrated with choroidal pigment, or there may be localised exudative deposits, or some clot remaining from blood effusion. In many cases no visible evidence of lesion can be detected, when the affection is probably of cerebral origin. Careful examination of the field of vision is the best guide to diagnosis in these latter instances. There will be a certain constant space or spaces in the field where impairment or complete loss of vision is noticed, and, in proportion to the nearness of the scotoma to the region of the macula lutea, by so much the more marked and obtrusive the defect will be, and, as a matter of course, the more unfit the patient for military service. If the impairment or loss of vision is confined to the scotoma, and the rest of the visual field remains normal after some months have elapsed, there is good reason for hoping the disease may not extend ; but when it is accompanied by more or less general loss of retinal power there is reason for fearing it will prove to be progressive until blindness results.

II. ASTHENOPIA.

Definition.—Deficiency of ocular strength, or feebleness of vision due to strained, unduly balanced, or irregular muscular action, altogether irrespective of amblyopia, which may or may not be present.

Causes.—Excessive strain, over-fatigue, or atony of the ciliary muscle. Weakness of the internal recti muscles relatively to the demands made on them for converging the eyes to near objects, as in reading. Want of perfect concurrence in the respective actions of the muscles concerned in accommodation and convergence. These causes of asthenopia are often, but not invariably, associated with ametropic states of ocular refraction, viz. hypermetropia, myopia, and astigmatism.

The symptoms which result from the two sources just named are in many respects similar, but, the causes being so different in their nature, the diagnosis between the different forms of asthenopia to which they respectively give rise should be clearly made out, for the principles on which their treatment is to be conducted must be equally different. Two forms of asthenopia are, therefore, distinguished, viz. (a) " motor asthenopia," and (b) " accommodatory asthenopia."

Symptoms.—The following symptoms are common to both forms of asthenopia. Reading and observation of near objects in general quickly induces fatigue. The effect of continued application of this kind causes a sense of fulness and tightness in the eyes, congestion, uneasiness about the brow, headache, and various forms of nervous disorder. On trying to read, the letters at first appear clear and distinct, but afterwards become blurred, and more or less cross each other (incomplete diplopia). Conjunctival vascular injection and epiphora follow if the exertion be continued. The symptoms are relieved by rest, and generally, according to the length of the interval of rest, so is the degree of relief.

Diagnosis between Motor Asthenopia and Accommodatory Asthenopia.—The power of mobility of the eyes must be examined, more especially the power of convergence of the optic axes. In *accommodatory asthenopia* mobility is unimpaired and the convergence of the optic axes perfect; in *motor asthenopia* the opposite condition exists. The following is a simple way of examining the amount of convergent power possessed by the asthenopic eyes. An object, such as a ruler, is held up before the face in a line midway between the eyes, about the distance of a foot off. This is slowly moved towards the face, and when only half a foot off attention is paid to ascertain if one of the eyes becomes unsteady and turns outwards. Should this happen after repeated observations, it shows that the internal rectus of the deviated eye is too weak to keep the eye in an inward direction. If the eyes be free from motor asthenopia, they will converge together symmetrically to the last limit of convergence. Another plan is to shade one eye, and to direct the other at some object; if *motor asthenopia* be present, the covered eye will be moved by the stronger external rectus muscle, and turned more or less outwards. The relative strength of the converging and diverging muscles may be determined by testing their power of counteracting the deflection of rays which is caused by applying prismatic glasses of known degrees of strength before the eyes. (See 'Prisms,' p. 32.)

The symptoms of motor asthenopia were attributed by Von Graefe to 'insufficiency of the recti interni muscles,' and he gave the following as a test of the precise amount of insufficiency. A prism with a refracting angle of about 12° is to be placed before one eye with its base horizontally downwards or upwards. The image of an object looked at by the two eyes will then be displaced upwards or downwards so far as the eye that has the prism before it is concerned. There will be vertical diplopia. If under these conditions a line with a dot marked upon it near its middle be looked at by both eyes at a distance of about a foot from the face, and a single line is still seen, though elongated, with two dots upon it, one above the other, no insufficiency is to be supposed to exist, as the internal straight muscles are obviously still acting in concord; but if two lines are seen, each with a dot upon it though at different elevations, the separation is regarded as a proof of insufficiency. Supposing the horizontal diplopia results from relative divergence of the optic axes, so that the two images are crossed, the distance at which the two lines are separated apart is taken as the indication of the amount of the insufficiency of the internal recti. The strength of the prism which with its base inwards would then produce fusion of the two lines gives the exact measure of the insufficiency. The accuracy of these views regarding 'muscular insufficiency' has, however, been frequently questioned, and other explanations, which certainly appear to be far more satisfactory, have been advanced to account for the facts observed in the experiments mentioned. It is not, however, necessary to enter more fully into the subject here.[1]

(a) *Motor Asthenopia* usually co-exists with *myopia*. From the closeness with which the myope regards objects, the internal recti m. are kept on a constant strain, and in certain cases this strain speedily induces exhaustion of muscular energy. There is not a corresponding amount of exertion of the accommodatory function, so the normal balance of action between the two functions of accommodation and convergence is broken. In high degrees of myopia, in young persons, there may be no demand at all for accommodation, while the strain of convergence is very considerable. When the distant point of distinct vision is only a few inches off, the myope will read and work at that distance, in which case no accommodation will have to be exerted, but for the sake of single vision there must be the necessary convergence. Exhaustion follows this excess of strain on the muscles of convergence, and the loss of the support that they should normally obtain from the harmonious co-operation of the function of accommodation favours its occurrence. If the patient, to counteract the effects of this exhaustion, increases his efforts to maintain the convergence of the eyes instead of giving them the necessary amount of rest, the pain and other symptoms of motor asthenopia follow. If neglected, this state of things may lead to permanent strabismus. The patient, in order to prevent the annoyance of the partial *diplopia* produced by the two eyes, owing to their unsteadiness, not seeing near objects precisely in the same direction, will use one eye only; the other will move outwards, and, if habitually unemployed, not only will squint be permanently established, but the eye will be rendered amblyopic. This has been more fully explained under Myopia, at p. 64.

(b) *Accommodatory Asthenopia.*—The cause of this form of asthenopia, namely, want of sufficient power in the muscle of accommodation to meet the demands made upon it, at once shows that whatever condition of the eye induces an excessive strain on this muscle, relatively to its general tone and development, must aggravate, even if it has not originated, the defect. Hence its

---

[1] Full explanations of the facts which were relied on by Von Graefe as proofs of insufficiency of the internal recti muscles, now generally acknowledged to have been erroneously relied on, may be found in the last of the admirable lectures which were delivered by Mr. Brudenell Carter at the Royal College of Surgeons, on 'Certain Defects of Vision.' These lectures were published by Macmillan & Co., London, 1877.

constant co-existence with *hypermetropia*, taxing so unceasingly, as this form of ametropia does, at all distances, the accommodatory function unless artificially relieved; its frequent occurrence also with unaided *presbyopia*, when work on small objects at near distances is persisted in; and its aggravation in degree if astigmatism, with the varying struggle of accommodation which it leads to, be superadded. There is also in accommodatory asthenopia a derangement of the normal co-operation of the functions of accommodation and convergence by which the asthenopia may be aggravated, but the disturbance arises from a different cause to that which originates it in motor asthenopia. In the latter form, as already mentioned, the action of convergence surpasses that of accommodation; but in accommodatory asthenopia, especially when it co-exists with H., the accommodation has to be exerted in excess of the convergence. Here also the loss of the support which springs from the concurrent and even working of the muscles concerned in accommodation and convergence no doubt augments the asthenopia and the difficulties of the patient. After he has been reading or writing for a time, or working intently at any close work, the objects looked at become indistinct, and if only stronger exertion is made to get clear images, instead of sufficiently resting the eyes, the symptoms elsewhere described are produced. These symptoms are aggravated by anything that deranges the general health and nervous system, or lowers the strength of the asthenopic patient. But there is not the tendency to eversion of the eyeball that there is in motor asthenopia. Moreover accommodatory asthenopia is relieved by convex glasses, in most cases is removed by them when they are of suitable strength, and these would only add to the difficulties of the motor asthenopia of myopes.

Loss of adequate power in the ciliary muscle and subjectiveness to asthenopia may be brought about by other conditions than those of hypermetropia, presbyopia, and astigmatism. Irritation from constant occupation at near objects, spasm, general debility, hysterical states, all become inducing causes of accommodatory asthenopia when the ciliary apparatus is subjected to efforts out of proportion to its strength.

Treatment of Asthenopia.—Asthenopia admits of treatment and often of cure. The ocular conditions which lead to its production point sufficiently to the means which should be adopted for its relief. If the eyes of the patient are found on examination to be ametropic, the kind of ametropia must be ascertained, and the defective refraction corrected in the manner already described in the chapters on myopia, hypermetropia, and astigmatism. If the patient has become presbyopic, relief must be afforded as explained under presbyopia. At the same time the relations between the convergence of the eyes and the accommodation must be considered, and as far as practicable a normal balance established between them. The employment of the eyes must be regulated, so that while, on the one hand, irritation and fatigue are obviated by the prevention of an excessive demand on the muscles of accommodation and convergence, especially under unsanitary conditions, such, for example, as sometimes are met with when printers are employed in setting up small types in close ill-ventilated rooms by gaslight, and by avoiding excessive straining of the eyes, as occurs when persons work at small objects, writing, drawing, &c., in tropical countries, with ill-judged exposure of the eyes to the direct solar glare; on the other hand, the weakening effects of abstaining from all proper employment of them are avoided. If there be weakness from constitutional disorder of any kind, its influence on the asthenopia of the patient should not be forgotten, but steps should be taken according to its nature to try and remove it, and to restore the natural standard of health.

III. Nephelopia.

Impaired Vision, or Dimsightedness, from Causes which obstruct the Passage of Light through the Eye to the Retina.—Nephelopia, cloudy or foggy vision, the result of lesions which have led to physical changes in the condition of some of the ocular media, especially to diminished transparency, is a frequent cause of rejection of recruits, and a source of unfitness for further military service among soldiers in the ranks. Such morbid changes, when they exist to an extent sufficient to cause complete incapacity for military service, are easily recognised, either by direct observation, lateral illumination, or ophthalmoscopic observation of the eye or eyes concerned; but in minor degrees, although impeding clear V., they are liable, without special care, to escape detection. The particular causes which lead to loss of transparency in one or more of the dioptric media, or otherwise obstruct vision, are as numerous as the disorders to which the various parts of the eye are subject. The consecutive effects of inflammatory lesions affect vision in very various ways and degrees. Slight interstitial haziness of the cornea may interfere with acuteness of vision by causing diffusion of the rays of light which traverse its substance, and so producing confusion of images on the retina; or the opacity may exist to such an extent, especially if it be central in position, as to obstruct the passage to the macula lutea of a large proportion of the luminous rays, and thus prevent all practically useful power of sight. Various morbid as well as senile changes in the crystalline lens and vitreous humour lead to similar results. Again the central parts of the dioptric media may be left moderately clear, but the iris may be more or less contracted and adherent, so that the pupil, owing to the smallness of its aperture, will only admit limited portions of the beams of light proceeding from illumined objects before the eye, and their brightness becomes lessened to such an extent as to prevent a proper view of them by the observer. This will especially occur when the subject of such a condition passes from a bright to a relatively dull light. Morbid states of the eyelids and other ocular appendages, and many other affections external to the globe of the eye as well as internal to it, which need not be mentioned here, will also lead indirectly to changes which interfere more or less with the visual function by obstructing the free passage of rays of light to the retina. It is with the permanent effects of these lesions and the influence they exert on sharpness of sight that the military surgeon has principally to deal, so far as optical fitness for military service is concerned; in their earlier stages such lesions are matters for surgical treatment.

CHAPTER VIII.

Regulations on Visual Examination of Recruits.—The instructions, dated 1st July, 1870, for the medical examination of recruits (clause D, para. 3), required that the recruit should be free from defects of vision—that 'he sees well.' It was further ordered in para. 8 of the same clause D that 'the special tests for power and range of vision are to be applied to each eye as directed on the card of test-dots furnished for that purpose.' The revised Army Medical Regulations of November, 1878, qualified in a certain degree and explained the foregoing requirement of the recruit 'seeing well,' inasmuch as, while directing by para. 559, p. 90, that men presenting 'defects of vision' will be rejected as recruits, it defined in para. 557 that one of the principal points to be attended to in the inspection of the recruit is 'that his vision is sufficiently good to enable him to see clearly with either eye at the required distance.' This direction, of course, implied that it was objects of a certain fixed size, viz. the test-dots, which were to be seen clearly at the distance required. The same directions are continued in paras. 969 and 970 of the Army Medical Regulations of 1885, while, in addition, in para. 986, the mode of testing vision by the test-dot card is fully described.

Under 'defects of vision' are obviously included all conditions of the eyes and their appendages which interfere with clearness of view of the appointed test-objects at the prescribed distance.   The necessity of a proper search for these defects is comprehended in the direction in para. 982, p. 171, of the regulations last referred to, viz. 'The surgeon examines the eyes and eyelids,' and is again alluded to in the succeeding para. 983.

Range of Vision necessary in Recruits.—The recruit is thus required to possess a certain range and degree of acuteness of vision ; and that he does possess this required range and power of vision is tested by means of the test-dots referred to in the regulations already quoted.   It is necessary to understand what range and power of vision are indicated by these test-dots, more especially as they have occasionally been supposed to be supplied for indicating that a recruit possesses a complete range and perfect acuteness or power of vision.

At first, when the introduction of long-range rifles with graduated aims in place of smooth-bore muskets made it necessary to pay particular attention to the range and acuteness of vision possessed by recruits, efforts were made to obtain recruits with full range and perfect acuteness of vision.   But it was found impossible to obtain recruits possessing such fine qualities of vision in sufficient numbers, and it therefore became necessary to relax the requirements in these respects.   It was then obviously necessary to have some standard range and power of sight fixed, such that the proved possession of them would render men acceptable as recruits so far as quality of vision was concerned, while the absence of them would render men unacceptable as recruits.

Ultimately a particular limit was determined, and this limit, which then became the regulation standard of visual fitness for army recruits, was published in the following order, a copy of which was issued to each army medical officer :—

Army Medical Department,
3rd December, 1863.

Sir,—His Royal Highness the Field Marshal Commanding in Chief having been pleased to notify
'That men should not be received into the service who do not see well, to 600 yards at least, a black centre 3 feet in diameter on a white ground,'
I have the honour to request you will have the goodness to pay strict attention to this command in the examination of recruits.

(Signed)   J. B. Gibson,
Director-General.

The black centre 3 feet in diameter on a white ground, mentioned in the foregoing circular, signified the bull's-eye of the target which was used at that time by trained soldiers in practising with the rifle at distances from 600 to 900 yards. The question then arose how medical officers were to carry out this instruction in examining recruits, there being many manifest difficulties in the way of ascertaining that men could see the actual bull's-eyes at the required distance under the conditions in which the examination of recruits is ordinarily conducted.

Test-dots for Military Purposes.—I had already suggested in an article in the Army Medical Reports for 1860 the use of black discs, formed on principles explained in the paper referred to, for the purpose above mentioned. I now prepared some of these discs on a card, so that when held at a given distance they formed retinal images of the same sizes as the bull's-eyes, 3 feet in diameter at 600 yards, and, having submitted them for approval, they were adopted for effecting the desired object. The size of each of these discs or test-dots was one-fifth of an inch in diameter, and, though they were distributed over the card at irregular distances apart, no two adjoining test-dots were placed with a less interval between them than one equal to their own diameter. As optical tests, the dots would have been capable of giving more accurate results if they had been all placed at equal distances apart, as explained at p. 161, but the purpose of disposing them irregularly was to afford means of counteracting attempts at deception or guesswork. The distance at which the test-dot card was arranged to be held was 10 feet. This was considered to be a convenient distance, and the diameter of the small test-dot was then found by a simple calculation of proportion, viz. as 600 yards : 3 feet :: 10 feet : $\frac{1}{5}$th of an inch. The visual angle of the dot $\frac{1}{5}$th of an inch in diameter at 10 feet being the same as that of the large bull's-eye 3 feet in diameter at 600 yards, or, in other words, the diameters of the two discs being seen under equal angles, and the two discs therefore being of the same apparent size, it followed, other conditions being alike, that if the recruit could distinguish clearly the small bull's-eyes at 10 feet distance, discerning, among others, the dots which were only one diameter apart, he could equally see the 3-foot bull's-eyes at 600 yards as required. It was only for this purpose that the test-dots were devised—to test the ability of the recruit to see the 3-foot bull's-eye at the prescribed distance, not to test the nature of any refractive or other visual defect which might in particular instances prevent the miniature bull's-eyes from being counted.

Counting Test-dots.— It was found by practical trials that recruits could not be relied on for counting correctly more than seven or eight of the discs at a time, even though they were all separately visible to them, and the small bull's-eyes, or test-dots, were therefore at first limited to this number. But it was found that the limit in number was made known to the recruits by the 'bringers,' so that the recruits, probably judging by the amount of test-dot card exposed, occasionally guessed the number submitted to them, although they did not see the dots distinctly.

To counteract this trick, a large number of dots was printed on the test-dot card, and they were so disposed that, by means of a covering card of a certain shape, which could be shifted into six positions in front of the test-dot card, twenty-five variations in the number and relative positions of the dots could be obtained without exposing more than seven or eight at a time. The test-dot card was ultimately adopted in this shape, directions for using it being printed on the back.

Change from Circular to Rectangular Bull's-eyes.—Subsequently the circular bull's-eyes and centres of the iron targets were changed to rectangular bull's-eyes and centres. This was not done for any purpose connected with eyesight in musketry practice, but simply in consequence of it being found more easy to paint the bull's-eyes and centres accurately upon the targets, either singly or in combination, when they were cast with vertical and horizontal lines marked upon them in small squares. In January, 1868, a corresponding change was directed to be made in the test-dots. It was ordered that the 2-foot square bull's-eye should be seen by recruits at a distance of 600 yards, the same distance as that at which, by the instructions of December 3, 1863, the circular 3-foot bull's-eyes had been ordered to be seen. In arranging the test-dots to comply with this order, it was found convenient to have the dots made one-fifth of an inch square in size. To apply them as tests for carrying out the order respecting range of vision, they had to be placed at a distance of 15 feet from the recruit. As before, the distance at which the test-dots were to be held was determined by a simple calculation of proportion, viz. 2 feet : 600 yards :: $\frac{1}{5}$th of an inch : 15 feet. In other respects the square test-dot cards were similar to the former round test-dot cards.

Return to Circular Bull's-eyes.—In March, 1876, the shapes of the bulls'-eyes and lines enclosing the centres on the service targets were again changed by general order. The rectangular outlines were discontinued, and the circular outlines were reverted to. Not long afterwards a corresponding change in form of the test-dots was adopted (W. O. Form I, 1220.) This card of circular test-dots forms the authorised test for service purposes in recruit examinations at the present date (1888).

Test-dots of W. O. Form 1233.—In 1875 a distinguishing War Office number was given to the cards of square test-dots, so as to include them in the list of Forms authorised for issue in the public service. They were marked W. O. Form 1233, and under this designation were referred to in the revised Army Medical Regulations of 1878.

From an optical point of view square bull's-eyes and centres are defective as means of measurement of the respective visual merits of marksmen. As the aim of the marksman is directed on a central point, and as that point becomes the anterior pole of the visual line, the merit of a particular shot can only be fairly tested by an estimate of the radial distance at which it has struck the target from the central point aimed at. It is evident that with square bull's-eyes and centres the shots of two men striking at the same distance from the central point may be differently estimated. If the shot of one has struck in the direction of the diagonal of the 2-foot square bull's-eye at a distance of 15 or 16 inches from the central point, it would count as a ' bull's-eye,' while another shot at precisely the same distance from the central point, but in a direction perpendicular to one of the sides, would be outside the bull's-eye and only count as a ' centre.'

Target-centres and Bull's-eyes.—In rifle drill instruction a distinction is made between a bull's-eye and a centre. The technical term ' centre ' might easily lead to misconstruction, as it does not occupy the position of the real centre of a target. The bull's eye in the range-targets hitherto in use has consisted of a black figure on a white ground, varying in size according to the distance at which the target has been placed, and according to the class of marksman firing at it. Outside the bull's-eye was a white space bounded by black lines. The space within these lines and between them and the bull's-eye was called the *centre*. Outside these lines was the remainder of the target, and in target practice the stroke of a bullet which neither hits the bull's-eye nor the centre, but hits the target beyond their limits, is spoken of as an ' outer.' When the rectangular bull's-eyes and centres were in use, the size of the bull's-eye aimed at by recruits and soldiers of the 1st Class was 2' × 3', used for distances varying from 450 to 800 yards ; for the 2nd Class, 2' × 2', for distances from 250 to 600 yards ; of the 3rd Class, 2' × 1', for distances from 50 to 300 yards. The size of the bull's-eye ordered to be adopted as the test for vision was therefore that of the bull's-eye used by marksmen of the 2nd Class. When the circular bull's-eyes and centres were re-introduced, the diameter of the bull's-eye for the 3rd Class was 1 foot, and was fired at by recruits from 100 to 200 yards, and by trained soldiers from 200 up to 300 yards ; of the 2nd Class was 2 feet in diameter, and was fired at by recruits from 300 to 400 yards, and by trained soldiers from 500 up to 600 yards ; of the 1st Class was 3 feet in diameter, and was fired at by recruits from 500 to 600 yards, and by trained soldiers from 700 to 800 yards (see Rifle Exercise and Musketry Instruction, 1879, Part VI, p. 245, &c.). The black circular test-dots used in trying visual power had reference, therefore, to the bull's-eyes used by marksmen of the 1st Class. Still more recently further changes have been made. By the latest regulations (Musketry Instruction, Provisional, 1884 ; and G. O. 38, March 1885), the bull's-eye on the 3rd Class target, which is fired at by recruits from 100 to 200 yards, is 12 inches in diameter; the bull's-eye of the 2nd Class target, fired at from 300 to 400 yards, is 2 feet in diameter; that of the 1st Class target, from 400 to 500 yards, is 3 feet in diameter. The 3rd Class target is fired at by trained soldiers in their annual course of practice at distances varying from 150 to 300 yards; the 2nd Class target, from 500 to 600 yards; and the 1st Class target from 700 to 800 yards.

Visual Angles subtended by the Target Bull's-eyes at different Distances.—There is only exceptional uniformity in respect to the visual angles which the bull's-eyes subtend at the different distances they are fired at. The 12-inch bull's eye at 100 yards has a visual angle of nearly 12' (11' 28''), and consequently the same bull's-eye at 200 yards, the 2' bull's-eye at 400 yards, and the 3' bull's-eye at 600 yards are all seen under equal angles of 5' 44''. The 1' bull's-eye at 300 yards, and the 2' bull's eye at 600 yards, are each seen under visual angles of 3' 50''. The 3' bull's-eye at 500 yards' distance has a visual angle of 6' 54'', while at 800 yards it has a visual angle of 4' 18''. The largest visual angle presented by a bull's-eye at any distance is the 1 foot bull's-eye at 100 yards' distance, viz. 11' 28'', and, therefore, having the largest apparent size, should, other things being alike, be the most distinctly visible to a marksman. As by Snellen's standard such plain black objects on a white ground ought to be seen by an eye of average normal acuteness of vision under a visual angle of one minute, or, by Burchardt's sight-tests, under a visual angle of 2·15 minutes, it follows that in a good light in the open air, if the atmosphere be clear, the tax on visual power in regarding the bull's-eyes at target practice as hitherto employed has been by no means a severe one.

Figure Targets for Range Practice.—A notable change was made in March 1885 in the forms of the targets used for range firing both by recruits and trained men at the annual courses of practice. 'Figure Targets' and 'Head and Shoulder Targets,' which were previously restricted to so-called 'Field Practices,' have now become ordinary targets for range practice. The bull's-eyes, and circles defining the centres, are retained, but are so painted as to be no longer visible to the men aiming at the targets, with one exception, which is in the case of recruits firing at the 3rd Class target, when the bull's-eye is ordered to be marked white. In all other cases a black figure representing the shadow of a man is to be painted on the target, and to be the object aimed at, the bull's-eyes being only marked in outline. This figure in each target is 6 feet in height, 2 feet across at the part representing the shoulders and upper part of the trunk, and 1 foot across in the parts representing the face and lower extremities. The bull's-eye and centre line are so marked that, although they are not distinguishable by the firer, they are visible to markers in the butts near the targets. The respective values of hits upon different parts of the targets remain equal to what they were when the bull's-eyes and centres were visible to the firers, a hit in the centre being valued at three-fourths of a hit in the bull's-eye, and a hit outside the centre as half the value of one on the bull's-eye. In certain practices hits on the figure only have a value, and the value is equal, whatever part of the figure may be struck. For range practice in the 3rd Class target only one figure is shown; in the 2nd Class target three figures, placed side by side; in the 1st Class target four figures; while in the target for volley and independent firing either six or eight figures are placed side by side, according to distance. It may be advantageous in some cases to be aware of the degree of visual acuteness which is required for these figures to be seen at the distances at which they have to be fired at, and this knowledge can be best obtained by ascertaining the visual angles under which they are presented to the sight of a firer.

Visual Angles of Target Figures at different Distances. — The target figures are placed for practice at distances which vary from 100 to 800 yards. The visual angles under which the bull's-eyes in the several classes of targets are presented to the firers have been already named. The visual angles which the 6 ft. figures subtend at the respective distances they are fired at are as follows :—

Distance.		Distance.	
100 yards	= 1° 8′ 16″	500 yards	= 0° 13′ 45″
150 ,,	0 45 50	600 ,,	0 11 28
200 ,,	0 34 26	700 ,,	0 9 50
300 ,,	0 23 14	800 ,,	0 8 36
400 ,,	0 17 11		

The visual angles formed by the breadth of the figure, both the broader and narrower parts, may be at once determined by the foregoing table, for the visual angles subtended by the part 2 feet across will be one-third, and by the part 1 foot across one-sixth of the dimensions of the visual angles under which the height of the figure is seen at the different distances specified in the table. In like manner, on three figures being joined together in the 2nd Class target, and four figures in the 1st Class target, the visual angles under which the combined broad parts of the joined figures are seen will be simple multiples of the visual angle subtended by the corresponding part in the single figure. In the 2nd Class target the width at this part will be trebled, and will be the same as the height, viz. 6 feet across; in the 1st Class target, in which four figures are placed side by side, the width across the body will be 8 feet, and the visual angle will be increased by an extent equal to one-third of the visual angle under which the height of the 6-foot figure is seen at equal distances.

An acquaintance with the sizes of the visual angles subtended by the figures on the targets at the various distances at which they are placed for range practice will enable a medical officer to determine how far any soldier, whose acuteness of vision for distant objects has been previously ascertained, is competent to distinguish them for practice as a marksman, light and other conditions being alike. The fact that the visual angle is greatest in the vertical direction in the 3rd Class target, while it is greatest in the horizontal direction in the 1st Class target, will also call attention to the influence that may be exerted on vision if the eye of the firer happen to be astigmatic in formation.

Figure Targets optically regarded.—It is obvious that from an optical point of view, and also as regards relative merits in respect to accuracy of aim, the figure targets in those practices in which a hit on any part of the figure has an equal value, while a hit any-where outside it is regarded as a miss, are open to the same objec-tions as the square bull's-eyes and centres. Optically regarded, they are by no means of equal value. A shot at the bottom of the figure, about 3 feet from the centre, cannot have the same optical value as a hit which may be under a foot in certain directions from the centre, and yet is not admitted to be a hit at all in musketry exercise. On the other hand, there are certain advan-tages in the use of the figures : whether single or in groups, they more closely resemble the objects a soldier would have to fire at in warfare than circular bull's-eyes and centres ; and the argument that a hit anywhere will cause a wound which would disable an enemy probably explains the fact of their being estimated at equal values in musketry practice, whatever their distance from the centre may be, so long as the hit is within the outline of the figure. The men under instruction are still taught to try and hit the centre of the object aimed at, on account of the margin this allows for variations in direction and elevation, and it is in accordance with this principle that the circles representing bull's-eyes and centres are retained, though they are only visible to the markers in the range practices. They enable the superior quality of the marksman to be shown, who, without seeing a distinct bull's-eye to aim at, can place his shot nearest to the centre of the figure or group of figures. Although certain advantages may attend the plan of attaching equal values to hits irrespective of distance from the centre of the object aimed at in some practices, it should not be forgotten that the principle of making no distinction in the value of the shots is not a right one, optically regarded, for it puts different degrees of accuracy of aim, and probably different qualities of sight, all on the same level. A man who only hits the target at the distance of the head or foot of the figure may not be able to fire more accurately owing to some visual defect, and, in conse-quence, may be but little to be depended upon as a practical marksman on other occasions of more importance, especially under different circumstances of illumination and contrasts of colour.

Test-dots introduced among Snellen's Test-types.—About the time that the circular test-dots were introduced, an edition of

Snellen's test-types was printed for distribution among the medical officers of the British army.[1] Dr. Snellen at my request introduced the test-dots among his test-types, and it may be observed that he placed above them the number 54, to indicate the number of Paris feet, equal to 57 English feet, at which the test-dots ought to be held for testing normal acuteness of vision. At this distance the test-dots would form a visual angle of the 60th part of a degree, which, as already explained, Dr. Snellen has taken as the smallest visual angle under which an object is visible by an eye possessing average normal acuteness of vision under ordinary conditions.

But, as all Dr. Snellen's types and figures are rectangular objects, while the test-dots introduced among them are circular, it follows that a difference of calculation was required for the test-dots. The difference between the area of a circle and the area of a square should have been taken into account. The area of a square to that of a circle is as $1 : 0.7854$, and, taking 54 Paris feet as the distance at which a rectangular object one-fifth of an inch square should be seen by norma'ly acute vision, a circular object one-fifth of an inch in diameter would, in proportion, be only visible by an eye of normally acute vision at a distance of a little under 43 Paris feet.

If all recruits could read, it would be far better to use types of definite sizes, such as Snellen's, for the examination of vision. The visual acuteness could be definitely registered, or the possession of any fixed standard of power of sight determined. But, as by the latest returns (Army Medical Report, 1887) there were still 9 per cent. of the recruits who sought enlistment unable to read, a simpler test, such as that of merely counting a few spots of certain size at a given distance, is rendered necessary.

Degree of Visual Acuteness shown by Counting the Circular Test-dots at 10 feet.—It has been mentioned that the circular test-dots one-fifth of an inch in diameter, when first introduced, were ordered to be held at 10 feet distance from the recruit, this distance, so far as concerned the production of the image on the retina, being equivalent to that of the bull's-eye 3 feet in diameter when seen, as ordered, at 600 yards. As already explained, similar test-dots held at the same distance are again employed for testing the vision of recruits, and form the authorised standard by which their acceptance or rejection is at present regulated. But it has been shown that under average normal acuteness of vision they should be seen at a distance of about 43 feet. Therefore, since $10 : 43 :: 1 : 4.3$, it follows that recruits accepted under the visual test just named are accepted with a fraction less than one-fourth of average normal acuteness of vision. In other words, if Snellen's test-types were used instead of the test-dots as the standard of visual sufficiency, a recruit would be accepted who·

---

[1] *Test-Types for the Determination of the Acuteness of Vision.* By H. Snellen, M D. Second edition, Utrecht, 1864.

could only read the 20 feet type at a little under 5 feet distance (4 feet 7½ inches) instead of the full distance of 20 feet.

Visual Acuteness shown by Counting the Circular Test-dots at 5 feet.—It is ordered in the Regulations for the Militia, 1883, p. 32, that a medical officer, in the examination of a militia recruit, is to ascertain 'that his vision is good, or at least sufficiently good to enable him with his right eye to discern objects clearly at not less than 300 yards;' and it is laid down in the directions on the back of the test-dot card in present use that the test-dots are to be counted by a militia recruit at a distance of 5 feet, in accordance with the requirement just named. It follows, therefore, that a militia recruit may be accepted who has only half the minimum allowed for a recruit of the regular army, or between ⅓th and ⅛th of normal acuteness of vision. If Snellen's types were used as the test, then a recruit would be accepted for the militia by the regulated standard who could only read the 20-foot type at a distance of 2 feet 4 inches, instead of the normal distance of 20 feet. This equally applies to recruits for all departmental corps.

Visual Acuteness shown by Counting the Square Test-dots at 15 feet.—The rectangular test-dots one-fifth of an inch square had not been introduced when the English edition of Snellen's test-types was published. Had they been, the figure 54 would have been rightly attached to them as showing the number of Paris feet at which they ought to be placed from an eye in proof of normal acuteness of vision. The order required the 2-foot square bull's-eye to be seen at 600 yards, which is the same as requiring the ⅕″ square test-dot to be seen at 15 feet; for 2′ : 600 yards : : 0·2 inch : 15 feet. But as for normal acuteness of vision by Snellen's standard they should be seen at 54 feet,[1] it follows that recruits were accepted when these dots were held at 15 feet, with less than one-third of normal V. (15 : 54 : : 1 : 3·6).

Reduction of Distance for the Square Test-dots by Order of August, 1870.—In August, 1870, the following circular modifying the standard of vision to be determined by the test-dots was issued :—

---

[1] *Calculation of Distance with respect to Visual Angle* —The distance at which an object, the measure of whose extreme limits is known, ought to be placed in order to subtend an angle of one minute may be roughly ascertained without difficulty, for the radius of the circle of which that measure forms part under the angle named will give very closely the distance required.

Thus, taking the square test-dot under notice :—If an object 0·2 of an inch in measure occupies 1 minute of a circle, 1 degree of the circle will be equal to 12 inches, and the circumference will be 360 feet. The diameter being equal to the circumference divided by 3·1416, the radius will be 57 English feet, omitting fractions. Therefore at 57 feet distance the visual angle of the 0·2 of an inch test-dot will be one minute. All the distances in the English edition of Snellen's test-types (1864) were stated in French feet, and, as the ratio of English to French feet is 46 : 49, so, omitting fractions, the 57 English feet would be equal to 54 French feet, the number stated above the test-dots in Snellen's tables.

A fact, easy to remember, is that a circular object, when removed to a distance equal to 57 times the measure of its diameter, is seen under a visual angle of one degree.

Recruiting.   Horse Guards, 3rd August, 18.0.

Circular Memorandum.

With reference to the instructions for the medical examination of recruits, dated July 1, 1870, it is notified that the medical officer will adhere strictly to the necessity that the vision of the recruit be sufficiently good to enable him to see clearly ; that paragraph 3 of clause D be carefully attended to, but that paragraph 8 of the same clause, as regards short sight, is so far modified that each test-dot on the card now required to be seen distinctly at 15 feet may, till further notice, be tested for a distance of 10 feet only.

CLEM. A. EDWARDS,
I. G. of Recruiting.

By this Order a further reduction in visual acuteness took place, for myopic recruits were to be taken who could recognise the test-dots at 10 feet only. The required standard of V. was thus lowered from $\frac{1}{3\cdot6}$ to $\frac{1}{5\cdot4}$ of average normal acuteness of vision.

Reduction of required Visual Acuteness limited to that caused by Myopia.—It should be noticed that the circular limited this depression of the standard to cases of myopia, and it threw on the medical officer the responsibility of distinguishing between defective vision due to short sight, and that resulting from other ocular abnormal conditions or disorders. The medical officer was directed to adhere strictly to the necessity of only accepting a recruit with vision *sufficiently good to enable him to see clearly* : merely the range of vision within which myopic recruits were required to see clearly was curtailed. At the present time no special limitation is in force as regards myopic relatively to other recruits. The existing regulations require that the circular test-dots shall be seen distinctly enough to be counted correctly at a distance of 10 feet by all recruits of the regular army alike, excepting those for departmental corps.

Quality of Vision tested by the Dots a Minimum Quality.—It is not to be forgotten that the examination carried out by the card of test-dots is for a minimum quality, and that an unknown number of recruits, probably a large proportion who are passed fit for service under its application, have a range of vision far beyond that which is indicated by the trial. The precise number of men admitted into the service who possess a farther range and higher degree of acuteness of vision than those demanded by the authorised test dot standard could only be ascertained by testing the full range and acuteness of vision of each individual who is enlisted.

Rules for the use of the Test-dots.—When employing the test-dots for trying the power and range of vision of recruits, it is important that the rules laid down for the manner of using them should be duly attended to. The following are the instructions printed on the back of the test-dot card of the present pattern (No. 27 | Gen. No. 4909.   June, 1885) :—

Mem. Each dot corresponds, at a distance of 10 feet, with a bull's-eye, 3 feet in diameter, at 600 yards. This is the range of vision required for recruits for the regular army except those for departmental corps.

Directions for using the 'Test-dot Card.'

1. Place the recruit with his back to the light, and hold the test-dot card perfectly upright in front of him, letting the light fall fully on the card.

2. Measure off with precision 10 feet in the case of a recruit for the regular army, and 5 feet in the case of a recruit for a departmental corps or the militia, the range of vision required for such corps being only 300 yards.

3. Examine each eye separately. The eye not under trial should be shaded by the hand of an assistant, who will take care not to press on the eyeball.

4. Expose some of the 'dots,' not more than 7 or 8 at a time, and desire the recruit to name their number and positions; vary the groups frequently to provide against deception.

The Test-dot Card must be kept perfectly clean.

CHAPTER IX.

Mode of Conducting Visual Examination—Appliances used in the Examination—The Examination Room—Each Eye to be separately Tested—Defective V. of either Eye causes Rejection of a Recruit—Pressure upon the Eye to be avoided—Application of Test-dots—Relative V. of Right and Left Eye—Procedure if Imposition be suspected—Secondary Inspection of Recruits—Field of V.—Visual Examination of Soldiers—Application of Snellen's Test-types—To ascertain the Sources of Defective V.—Preliminary Inspection of the Front of the Eye—Lateral Illumination—Detection of M., H., and Ast.—Ambl., how to distinguish from M.—Ambl. complicated with M.—Disqualifying Degree of M., H., Ast., and Ambl. - Weak V. from other Causes—Colour-blindness—Degree of M. which does not disqualify for Service—Visual Quality essential for Soldiers—For different Parts of an Army—Degree of M. which disqualifies for Service—Myopic V. of $\frac{1}{23}$. or 1·75 D.—M. = $\frac{1}{12}$ or 1·25 D.—Ametropia in Continental Armies—Spectacles not worn by Soldiers in the Ranks—Spectacles at Musketry Instruction—V. necessary for Commissioned Officers—Declaration concerning V. by Candidates for Commissions—Blindness of One Eye—Quality of V. required for the Line—For the Indian Govt. Services—Attempts at Deception—Quality of V. required for Army and Indian Medical Staff—For the Royal Artillery and Engineers—For Special Instruction at School of Musketry—For Royal and Indian Navy—Impaired Vision of one Eye in a Soldier not a Cause for Discharge—Aiming with the Left Eye—Detection of Simulated Impairment of V.—Assumed Blindness of one Eye—Modes of Detection—Case in Illustration—Assumed Defective V. of both Eyes—Modes of Detection—Case in Illustration—Simulation of Defective V. seldom attempted by British Soldiers.

1. Visual Examination of Recruits.—The visual examination of a recruit by the test-dots need not occupy under ordinary circumstances more than a few seconds of time. When it is considered necessary, for special reasons, to determine further the quality of vision of a recruit, as when the evidence afforded by the test-dots leaves some ground for doubting its accuracy, the additional examination may perhaps occupy 10 or 15 minutes, and it is best. therefore, to allow the recruit to dress himself before this further examination is begun.

In the latest edition of the Medical Regulations (1885, p. 169) the recruit's vision is ordered' to be tried by the test-dots before proceeding to the general examination while he is undressed, probably because further examination is held to be unnecessary if he is obviously unable to comply with the tests submitted to him.

2. **Appliances for Visual Examination.**—If a particular examination of the quality of V. of a recruit is required, it must be made in the same manner as the visual examination of a soldier who is already in military service. The ordinary ophthalmoscopic and optical case is sufficient, in conjunction with the test-types and dots, for proving the quality of V. possessed by a soldier in all ordinary instances. For special purposes a complete case of trial lenses is necessary, and a good perimeter is desirable. Full sets of lenses afford facilities that cannot be obtained by any other means for solving complicated optical defects, and for sifting suspected cases of simulation. They are, therefore, especially useful in general and invaliding hospitals, to which such cases are commonly sent for decision. They also afford the means of proving, by positive correction of the defects, the correctness of estimates of abnormalities of refraction, and of amounts of loss of accommodatory power, which have been previously diagnosed elsewhere.

3. **Examination Room.**—Rooms in which the examination of recruits is conducted should be well and equally lighted in all directions. This arrangement is especially important in testing vision. The light falling on the test-objects to be looked at should resemble the ordinary external diffused daylight as nearly as possible. It should neither be dazzling nor obscure. If the examiner's V. is of normal standard, or its degree is known to him, the sufficiency of the available light can readily be tested by personal observation. The eyes of the men to be examined should be habituated to the degree of light in the room for some minutes before the examination is made. The conclusions would not be accurate if the tests were suddenly applied to a man who had just left a place that was either much darker or more brilliantly lighted than the examination room.

It is a matter of convenience and also a means of saving time to have some lines, showing distances in feet, or in metres and parts of metres, permanently marked in ink upon the floor of the room in which Snellen's types or the miniature bull's-eyes are used. Whenever the space is available the distance should extend at least to 20 feet or to 6 metres. The addition of a simple stand for suspending a table of Snellen's types at the end of the range thus marked out is also useful. They should be suspended on a level with the eyes of the person to be examined.

In conducting the examination, the soldier or recruit is placed with his back towards the window or source of light, so that while his eyes are in relative shade there is a full illumination of the types or dots by which his sight is tested. This is important as regards both the large and small types when test-types are used, but particularly so as regards the small types for trying near vision.

A range of 6 metres, or 20 feet, is essential for applying the most recent rules for testing the vision of candidates for commis- sions in the army, and, as described in the preceding paragraph, this range should be suitably placed in relation to the source of light.

4. **Separate Examination of each Eye.**—It is always necessary to test each eye separately. It will not often be found, even under healthy conditions, that the absolute refractive qualities of the two eyes of the same person, independent of accommodatory exertion, are precisely alike; but in defective conditions of vision the differ- ence between the two eyes is usually more marked. It has some- times occurred that a man has been blind in one eye without knowing it until attention has been directed to each eye separately by optical examination.

5. **Defect of Vision in either Eye of a Recruit.**—Under existing rules the existence of defective vision in either eye is a cause of medical rejection of a recruit seeking enlistment. Though the right eye may be up to the standard, if the vision of the left eye be so defective as to prevent the recruit from being able to count the test-dots with it, the orders are that he must be rejected. He must see clearly with each eye separately at the regulated distance. In the former days of long service with the colours the chances of the sight of the normal and efficient eye becoming independently affected by disease originating in the exposure and other causes incidental to military service, and of the man thus becoming dis- abled for duty and entitled to claim a pension, doubtless had an influence in determining the rule that not only the right eye, but the left also, of a recruit should be ascertained to be up to the authorised standard of visual acuteness before he is passed fit for acceptance as a soldier.

6. **Pressure on the Eye.**—In examining the eyes separately, an assistant should cover the eye not occupied in regarding the object, and not the man himself. If the man be allowed to close it he will probably, from carelessness or nervousness, exert undue pres- sure on the globe, disturb its condition for clear vision, and cause delay until this disturbance is recovered from. The assistant who covers the eyes should be taught that if any pressure be made it should be limited to the margin of the orbit. The object is simply to exclude light by closing the lids or shading the light with the hand; the eye itself should not be pressed upon. If undue

pressure have been made, the man's vision will be rendered misty, and it will be necessary to wait a minute or two until the eye has recovered its normal state before applying the test for visual power. The assistant should be taught to cover the eye by the palm of the hand, formed into a hollow for the purpose. If the fingers are employed, they are apt to press upon the globe, and there may inadvertently be vacant spaces left between them; in which case, either accidentally, or intentionally taking advantage of the opportunity, the eye not under examination may look through one or other of these openings.

If a trial case of lenses be available, it is still better to use the spectacle frame, and to place either a ground glass or opaque metal disc, both of which are usually supplied with such cases, in front of the eye not under trial. Not only is pressure on the globe thus prevented, and all chance of the person seeing through chinks obviated, but the freedom of movement of the eyelids is not interfered with.

7. **Application of Test-dots.**—If the man under examination be a recruit, as soon as he is placed in position, the test-dots are held upright before him at the prescribed distance—10 feet in the case of a recruit for the regular army, and 5 feet for militia and departmental corps recruits—and he is asked to state the number of dots exposed to his view, in the manner already explained at page 189. He should be required to count two or three series of dots with each eye, and if he replies readily and satisfactorily, so far as power of vision is concerned, he is fit for service.[1]

8. **Rejection of a Recruit.**—If the recruit should make repeated mistakes in counting the number of dots presented to him at the prescribed distance, and there is no reason for suspecting that he is not doing his best to try to see them clearly, especially if he should succeed in counting them correctly when they are held at some point nearer to him than the prescribed distance, he is then rejected as unfit for service on account of 'defective vision.' The regulations do not require that the nature or degree of the defect should be particularly stated at the first inspection of recruits.

9. **Relative Visual Acuteness of the two Eyes.**—As recruits in aiming and firing practices are taught to use the right eye only, the left being closed, it seems worth consideration whether it is necessary to insist on men seeking enlistment having the same range and power of sight in the left as in the right eye, particularly under a system of engagement for short service. Compliance with

---

[1] The following regulation on the application of test-dots is given in paragraph 986 of the latest published Army Medical Regulations of 1885. 'In examining a recruit's vision he will be placed with his back to the light, and made to count the dots and describe their position at the distances specified on the test-dot card, first with both eyes and then with each separately: the medical officer will manipulate the card, while the assistant covers each eye alternately with the flat of his hand.

the authorised standard of visual power does not exclude a man who has simple M. of about 1·5 D, or $\frac{1}{27}$. And, supposing this to be the degree of M. of his right eye, although the M. were as high as 3 D, $\frac{1}{13}$, in his left eye, there hardly seems to be any valid reason against accepting him as a recruit. Even if the visual acuteness of the left eye were lowered by the M. being complicated with Ast., so long as the visual defect is one of refraction only, not one of any morbid origin involving a liability to recurrence of the disorder, it would not prevent him from seeing pretty well for certain distances, and from doing ordinary duties in the ranks and in marching ; and there hardly seems to be good ground, therefore, under the present system of using the rifle, for rejecting him as a recruit if he be eligible in all other respects.

The quality of vision of the left eye seems to be regarded as of minor importance in the case of recruits for the militia, for it is specified, with regard to the medical examination, that one of the principal points to be attended to is that the recruit's vision is good, or at least sufficiently good to enable him with his *right* eye to discern objects clearly at not less than 300 yards, while no quality of vision is referred to as regards the left eye.—(Regs. for the Militia, W.O. 1883, Part I. sec. 2, p. 32.)

10. **Defective V. due to Nephelopia or opacity of some of the Anterior Dioptric Media.**—Diminished translucency of the central portions of the cornea consequent on keratitis, diffused deposit of lymph on the capsule of the crystalline lens after iritis, and, indeed, cloudiness of the anterior media of the eye from any cause, even though they may not exist to such an extent as to be readily obvious to external observation, cause indistinctness of vision from the diffusion of the rays of light by which the structures concerned happen to be traversed. If the loss of transparency is not considerable, the circular test-dots in present use may be counted at a distance of 10 feet in a good light, but it may be found that a recruit cannot count them beyond that distance. It is questionable, unless there is an urgent demand for recruits, whether a man should be regarded as fit for military service under such conditions, especially if the right eye is concerned, although he is able to count the test-dots at the prescribed distance. It is not merely that a maximum limit of 10 feet in counting the test-dots represents rather less than one-fourth of Snellen's standard of visual acuteness, but in a large proportion of such cases a liability to recurrence of inflammatory action exists, and should this take place an aggravation of the existing defect is almost inevitable.

11. **Procedure when Imposition is suspected.**—In the British service, as recruits are for the most part men voluntarily seeking enlistment as soldiers, not like the majority of conscripts in Continental armies trying to escape enlistment, if any efforts at all are made by them to practise imposition, they will probably be directed

o

to the concealment of any defects of V. they may labour under, rather than to their exaggeration.

If, however, there is cause for suspecting that the man who has volunteered for enlistment has changed his desire on the subject, and that he is trying to escape from the bargain he has so far entered into by assuming a visual defect which does not really exist, he must be subjected to further tests before he is pronounced unfit for service on this account.

12. **Visual Examination at Secondary Inspection of Recruits.**— Many instances have occurred in which a recruit at first examination has been passed fit in respect to visual power and range, but, on being subjected to secondary inspection, has been found unable to count the test-dots accurately at the proper distance with one or other eye, or even with both eyes. At the secondary inspection, when the test-dots are placed before the man, he may perhaps give a succession of wrong replies as to the number of dots exposed to his view. Great caution should be exercised in such a case as to the rejection of a recruit. The presumption is certainly that he was properly tested at the first examination, and that for private reasons he does not choose to count the test-dots correctly at the secondary inspection. Unless there is some evident ocular defect which has been overlooked at the former inspection to substantiate the man's statement, a case of this kind is calculated to excite strong suspicion of the disability being feigned, and systematic steps should be taken to ascertain whether the man's statements are true or false. This can be done by ascertaining the man's alleged V. when the test-dots are placed before him, and then comparing it with the results obtained from the application of other test objects, Snellen's types, for example, as elsewhere explained.

The following special direction on the point just referred to is published in the Army Medical Regulations of 1885, para. 987:— ' A recruit whose vision has been tested and pronounced good on a primary examination will not, through his own declared inability to see the test-dots on secondary examination, be rejected unless the approving medical officer is satisfied that the man's vision is really defective and no deception is being practised by him.'

13. **Field of Vision.**—It is important that a recruit should not only possess sufficient central acuteness of vision, but that he should also have his field of vision complete in both eyes. Loss of the outer or temporal portion of the field of view of either eye from any cause especially unfits a man for military duty, for it disables him from noticing the objects by which he may be surrounded on the defective side and, consequently, from properly steering his way among them. A considerable portion of the lateral field of view belongs solely to the eye on the side concerned, as elsewhere explained, and a deficiency in it is not in any degree supplied or

compensated for by the eye on the opposite side. The temporal portion is not a part of the field of view which is common to both eyes, and when this portion is absent, objects on the side concerned do not attract observation or attention so long as a soldier. is marching and looking directly forward. The manner in which the existence of deficiency in part of the field of vision is to be ascertained has been already described (see page 12).

14. **Visual Examination of Trained Soldiers.**—When it is necessary to determine the V. of a soldier already in the service, the general manner of conducting the examination is the same as with the recruit, only the test-types should be used instead of the test-dots. The trial by the test-dots at a prescribed distance, as before explained, has been specially ordered for men seeking admission into the army. In the cases of trained soldiers already serving, a complete examination is required, as more important issues depend on the decision at which the surgeon may arrive. The question usually submitted to the medical officer is whether the inefficiency of a soldier who shows himself to be a specially bad shot at particular distances at the annual course of musketry instruction, or who shows himself incapable of judging distance with an approach to correctness, is due to some visual defect or not. The nature of the defect, if defect exist, and the extent to which it disqualifies the man for duty, have therefore to be ascertained and stated with accuracy.

15. **Use of Snellen's Test-types.**—In such cases the degree of V. must first be determined, and this is very easily done by Snellen's test-types or test-figures. The mode of ascertaining and expressing the acuteness of vision by these objects has been explained at page 163.

16. **Vision not defective.**—If the tests are so answered as to show that the man under examination possesses average normal V., or a near approach to the average, all morbid states of the eye, such as limitation of the field of vision, as well as disorders of its appendages, being understood to be excluded, the subjects of complaint are manifestly not due to visual defect.

17. **Procedure if Vision be defective.**—If V. is proved to be considerably under the average, the cause of the deficiency must be ascertained.

In the first instance, particular and special attention must be paid to the examination of the anterior ocular structures. The cornea, anterior chamber, iris, and crystalline lens of the affected eye should be subjected to minute observation. In the case of a recruit at secondary inspection, defects may possibly be found to exist in these structures sufficient partially to obscure V. which were not perceived in the observation of the eyes made at the primary inspection, although this could hardly occur if the tests for eyesight had been properly applied. They may equally exist in the case of a soldier serving in the ranks, without being visible by ordinary observation, as the result of some inflammatory action to which the eyes have been subjected. This preliminary inspection is important and should invariably be made. Considerable time is often wasted afterwards when it has been neglected.

18. **Lateral Illumination.**—The superficial examination is very rapidly made by *lateral illumination*, and indeed can only be thoroughly accomplished by its means. Lateral illumination signifies lighting up the parts required to be observed by concentrating upon them a pencil of rays cast in an oblique direction. The man is brought near the window of the room, one of the bi-convex object lenses in the optical case is placed vertically near the outer angle of the eye under observation in such a way that the light passing through it is made to converge upon the cornea, or through the

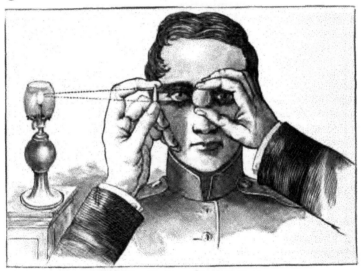

FIG. 74.—LATERAL ILLUMINATION.

cornea on the iris or lens, and the condition of all these structures is then examined by the spectator standing in front. This lateral illumination is of course more brilliantly seen when the flame in the ophthalmoscopic room is used as the source of light, but is sufficiently marked by solar light on any ordinarily clear day. By these means the slightest roughness of the surface of the cornea, interstitial haziness, minute ulcers, or the remains of them, fine exudations at the margin of the pupil, posterior synechiæ, commencing cataract, are made most obvious to observation, while the pencil of rays at or near its focus is made to play upon each structure at pleasure. Nothing can be more beautiful than the perfect precision with which opacities of the cornea and lens, adhesions of lymph to the capsule, and other morbid changes, some of which may not be perceptible under ordinary observation, become defined by light thrown laterally upon them in the manner just described. If minuter observation be required, the objects while thus illuminated by a lens held in one hand may be magnified by a second lens held within its focal distance by the other hand in front of the eye, without impairing their brilliancy or distinctness of outline (see fig. 74). A little practice will enable the operator at one and the same time to direct the apex of the luminous cone proceeding from the lens held between the thumb and forefinger of one hand upon any point of the anterior segment of the eye at pleasure, while he so adjusts the lens held between the thumb and forefinger of the other hand in front of the eye as to obtain the enlarged view of the parts thus illuminated which he wishes to examine more particularly.

A sufficient explanation of the impairment of sight will sometimes be found in this preliminary examination of the eye; but if nothing abnormal can be thus detected, the examiner must proceed further, with a view to discovering the source of the defective vision under which the recruit or soldier appears to labour. It may be due to ametropia, astigmatism, amblyopia, or disease of some of the structures constituting the fundus of the eye.

19. **Detection of Myopia.**—The recruit who has not been able to count the test-dots at the distance of 10 feet is found able on examination to count them at some distance short of it. Equally the soldier who has not been able to read the 20-ft. or 6-m. types at the proper distances is able to read them at some fixed distance nearer to the eye. In both instances V. is much improved, and the dots and types distinguished farther away by the application of a stenopœic hole. The Mc. soldier is found able to read the small No. 1 or 0·5 types at or near the distance of 12 or 20 inches respectively from the eye. The external signs described in the general remarks on M. will probably at once cause the surgeon to judge that no simulation is being practised, and indicate the true nature of the affection he has to deal with. The position of the near point of V., in relation to age, will help in the diagnosis, and, further, the use of the convex spectacles, as explained in the section on Myopia, will tend to remove any doubts that may remain. But the refractive condition of the eye should always be objectively determined by the ophthalmoscope in suspicious cases, in the manner described in the third chapter of this work. This applies to every case of suspected exaggeration or simulation of Mc. defect, for it enables the surgeon to arrive at a conclusion on the state of refraction, which it is out of the power of the man himself to control.

20. **Detection of Hypermetropia.**—The hesitation and difficulty which the Hc. recruit or soldier exhibits in recognising the types and dots at any distance, and the form of the eye, will probably lead the surgeon to suspect the affection under which the man is labouring. If, in addition to the largest types being only read

with effort, the smallest (No. 1) type cannot be recognised, or only with much difficulty, while the No. 2 or the 0·5 types appear more distinct at a distance than when brought close to the eye, not only H. but a deficiency of Acc. for near V. are indicated. Here, again, the position of the near point of V., relatively to its normal position according to age, should be noted. But the diagnosis should be established objectively by the ophthalmoscope as well as by the convex spectacles, as already explained in the section on H.

21. **Detection of Astigmatism.**—If the man under examination has exhibit.·d hesitation or particular difficulty in recognising the letters or counting the dots at any distance, and the tests for simple H. and M. are not readily responded to, the existence of Ast. may be suspected. The man should be then tried by Snellen's 20-ft. vertical and horizontal lines or dots, or the astigmatic Fan. He should also be subjected to ophthalmoscopic examination in order that the diagnosis may be established in the manner explained in a previous part of this work. The method of proceeding for determining the kind and degree of Ast., supposing its presence to be established, has been explained in the special section on Astigmatism.

22. **To distinguish Amblyopia from M.**—If, although the transparency of the ocular media be unimpaired, it be found that the smallest-sized types and dots cannot be distinctly seen at any distance, and that larger-sized type is held nearer than normal to the eye, the recruit is probably labouring under *amblyopia* ; he cannot be affected with simple *myopia*. To ascertain if this be the cause of his defective vision, weak concave lenses, when these are at hand, are placed before his eyes. If he now sees distant objects worse than he did before, his defect is almost beyond doubt some form of Ambl.; for the concave glasses which would improve vision, if he were myopic, render vision worse in Ambl. by diminishing the retinal images. When the convex 4 D spectacles are placed before his eyes, if the man under examination be emmetropic but at the same time amblyopic, he will be able to see the 1·5 D type, or type of larger size, at a distance of 10″ from the lenses, although he is unable to see the smaller-sized types at that distance.

The stenopœic hole may also be used for settling the question. If a man be Mc. and look through the stenopœic aperture in a metal diaphragm, or through a pinhole in a card at distant objects, he will perceive them far more clearly ; while if he be amblyopic, there will not only be no improvement, but the objects will appear more obscure. A person affected with a high degree of M., as M.$=4$ D or 5 D, will probably be able to recognise Snellen's moderately sized types (D$=2$ or 2·25) at double the distance when regarding them through a stenopœic hole that he will be able to do with his naked eye, while a person affected with Ambl. will not be able to see them any farther, and probably not so far off, by this proceeding.

23. **Amblyopia complicated with M.**—M. may be complicated with Ambl., and it is important to distinguish simple short-sightedness from short-sightedness with this complication. In the latter case, when the $+4$ D spectacles are worn, the subject will see types up to some distance within 10 inches—say, for example, 5 inches. The patient's true far point of distinct V. is then at 10 inches, and if the M. be uncomplicated with Ambl., he will be able to read the 0·5 D type at the distance of 10 inches without the aid of lenses. If it be complicated with Ambl. he will require the 0·5 D type to be held nearer to the eye than 10 inches; he will be only able to read some of the larger types at that distance. Again, simple M. is completely corrected by suitable lenses, and hence a ready mode of establishing an exact diagnosis between simple M. and M. with Ambl. is afforded. Having proved the existence of M. according to the methods already explained, the degree of M. is next determined. The proper concave lenses to correct this degree are then applied, and with them the simply myopic person will be able to see plainly the types of various sizes at their proper distances; but if he be also amblyopic he will only be able to see types of larger sizes at them, according to the degree of Ambl. By referring to the description of the causes of Ambl. it will be seen that the occurrence of Ambl., if extensive, with M. is a grave complication : for it indicates the existence of disease, and not merely peculiar conformation of the eye. The nature of this disease must be solved by ophthalmoscopic examination.

24. **Regulated Conditions under which M. disqualifies for Military Service.**—If the test-dots cannot be counted 10 feet off, owing to M., the man under trial is to be rejected as a recruit. Myopia, therefore, when accompanied with V. inferior to $\frac{1}{4\cdot3}$ Sn., disqualifies for military service. No special degree of M. has been fixed on as disqualifying for military service, but the disqualifying degree may be approximately estimated from the extent to which its interference with acuteness of vision is held to unfit a man for military service. This is considered farther on.

25. **Condition under which H. disqualifies.**—The recruit who cannot count the test-dots at 10-ft. distance, and whose inability to do so is proved to be due to H., by existing rules is disqualified for service. Hypermetropia, accompanied with V. inferior to $\frac{1}{4\cdot3}$, therefore excludes from enlistment as a recruit. It must not be forgotten that the disabling effects of H. are more or less felt in both far and near vision, that they increase with age and with diminished power of exerting Acc. from any cause, and that they are sure to be aggravated by anything that weakens the soldier.

26. **Conditions under which Ast. disqualifies.**—The same rule applies when the eye is found to be astigmatic. Ast., therefore,

of any form, when it is accompanied with V. inferior to $\frac{1}{4 \cdot 3}$, excludes from enlistment as a recruit.

27. **Disqualifying Degree of Amblyopia.**—The general rule that the test-dots must be counted readily at 10-ft. distance by each eye determines the disqualifying degree of amblyopia. Whatever may be the cause of the Ambl., if it exist to an extent which prevents a recruit from counting the test-dots at a distance of 10 feet from the eye, it disqualifies the man for military service. It is not ordered that the cause of the Ambl. is to be determined in the case of a recruit. The reduction of V. to about one-fourth of the normal standard from Ambl. is more disabling than a corresponding reduction owing to M. or H., for the defect in the case of Ambl. does not admit of correction or amelioration by lenses.

28. **Disqualifying Degree of Nephelopia.**—Loss of transparency in some parts of the dioptric media from previous inflammatory action will be found in most of the instances comprised under this heading; but whatever the visual defect may be in a given case, or whatever may have been the pathological conditions or lesions which have led to the dim-sightedness, the fitness or unfitness for military service of the subject of it, according to existing rules, is tested in the same manner as the degree of V. in other instances If the affection do not hinder a recruit from counting the test-dots at the fixed distance of 10 feet, it does not exclude him from military service; but if V. is so reduced as to prevent him from counting the test-dots at that distance he is ineligible.

If the nature of the affection should be such as to indicate a liability to recurrence of the morbid action which originated it, in which case an aggravation of the defect might be anticipated, even though the man could pass the test-dot examination, it would manifestly, under the circumstances, be imprudent to declare the subject fit for military service; while the same condition in a trained soldier would not justify his discharge from the service on the mere assumption of a similar tendency.

29. **Disqualifying Colour-blindness.**—Colour-blindness is not specially named in any military regulations as a ground for rejecting a recruit for the ranks, and recruits are not usually tested in respect to sense of colour. It might, however, if attention were directed to the subject, be comprehended in the general term ' defects of vision.'

All candidates for employment in the Royal Navy, officers and men, are tested with regard to their capability of distinguishing colours, and if they are discovered to be colour-blind, the defect is a cause of rejection. They come under the regulation, quoted on p. 151, which is laid down in para. 1,074 of the Queen's Regulations and Admiralty Instructions. This rule includes medical as well as other officers. It is obviously necessary that medical officers should possess normal colour-perception, as they are the usual examiners for colour-sense in others.   The conditions which render normal perception of colour in the officers and men of the Royal Navy so important have been already described at p. 152, in Chapter VI.

30. **Quality of Vision essential for Soldiers in the Ranks.**— Whatever visual disorder a non-commissioned officer or soldier serving in the ranks may happen to contract, so long as he is left with V. equal to that for which recruits are tested, he must be regarded as visually qualified for military duty.   If he can read the 6-metre type of Sn. at a distance of 1 metre and a half, or count the army test dots at 10 feet, he has still the power of sight which would have sufficed for his admission into the military service.

The remarks regarding the need of a complete field of vision for a recruit equally apply to soldiers in the ranks.   Hemiopia, and especially loss of the temporal portion of the field of view, due to causes occurring subsequent to enlistment, afford sufficient ground for the subject being brought forward for discharge from further military service.

31. **Quality of Vision essential for a Marksman at full Rifle Range.** —The capacity for becoming a perfectly reliable marksman at all the distances for which a modern rifle can be adjusted implies the possession of normal acuteness of vision in the right, or aiming eye, from the natural near point of distinct vision up to the remotest distance; and this faculty can only exist when the eye is Emc. or so nearly so as not to exceed 0·5 D of ametropia, and when the Acc. is also normal.   At the same time the left eye, or eye not employed in taking aim, ought not to possess less than three-fourths of the normal standard of visual acuteness.

32. **Degree of M. permissible in Recruits by existing Orders.**—As M. is by no means an uncommon affection, though far from being as common as it is in some foreign countries, it becomes important to be aware of the degree which, according to existing regulations, allows a recruit to be passed as fit for military service in the combatant ranks.   Uncorrected M. in a soldier is a grave matter, not only on account of its incapacitating him for the accurate use of his rifle, but also because it may lead to the safety of an important post, which he has been placed on sentry to guard, becoming endangered owing to his limited range of distinct view, especially as daylight diminishes.   No set limit has been defined with respect to the degree of M. which incapacitates for service in the English army; but it may approximately be arrived at by experimental observation of the degree of uncomplicated M. which admits of the test-dots being counted at the distance, 10 feet, which determines recruits to be eligible for service so far as vision is concerned.

Experimental trials show that persons of equal ages, and still more persons of varying ages, differ considerably in their power of distinguishing the presence of objects, notwithstanding that the retinal images of the objects are obscured by an equal amount of blurring from diffusion owing to similar myopic conformation of the eyes. This circumstance results from the fact that absolutely perfect definition of a retinal image is not necessary for recognition of an object. There are few, if any, eyes which can see a star as the luminous orb which it is. All the other conditions, too, on which acuteness of vision depends, besides perfect definition of retinal images, are subject to variations in different individuals. Differences exist in the power of preventing peripheral rays from entering the eyes as well as of eliminating the confusing effects of diffusion of retinal images, in habits of observation of objects, quickness of perception, and other influences of the kind. But taking the average of a number of trials, at about the ordinary ages of recruits, I have found that persons affected with uncomplicated $M. = 1\cdot75$ D can manage, with each eye singly, to count the test-dots at 10 feet under suitable exposure in good daylight. The test-dots are seen mistily; they may appear more or less altered in form, but they can be sufficiently distinguished to admit of being counted. The present test-dot standard for vision of recruits, therefore, admits men with an amount of $M. = 1\cdot75$ D, and with visual acuteness less than $\frac{1}{4}$ Sn. ($\frac{1}{4}$). It will usually exclude degrees of M. higher than $1\cdot75$ D.

33. **The Qualities of V. which are needed in different Parts of an Army.**—The M. which would make a man unsuitable for the duties of one arm of the service may not make him unsuitable for another. The M. which would unfit a soldier for aiming at long ranges, whether with a rifle or a field gun, or for the duties of a cavalry vidette, would not unfit him for the working duties of a sapper or pioneer, or for those of the commissariat and transport corps, or medical staff corps. Just as there are different standards of height, girth of chest, &c., for the men of different parts of the army, so equally necessary appear to be different standards of V. to fit them for their special duties. Certainly riflemen, artillerists, and cavalry soldiers, ' the eyes of the army,' should especially be as free as possible from short-sightedness, hypermetropia, and other defects of vision.

What particular degrees of M. and H., however, should exclude men from special parts of an army in which very acute vision, and a long visual range, are essentially important, can only be determined after a definition by military authority of the exact requirements in those several parts of the service. It is the province of the military authorities to settle the degrees of V. which are necessary for the military duties and responsibilities demanded from soldiers in each branch of the military service; it is the province of the medical officer to ascertain that the men possess the precise degree of V. which is ordered. At present, in the enlistment of recruits for the British army, the same orders in respect to the examination of vision hold good for all recruits alike, whatever branch of the regular army they may be destined for.

34. Degree of M. which unfits for Military Duties in the Ranks.—The circumstances of military service are so different in the British army from what they are in continental armies, the cost as regards the individual soldier is so much greater, that in the selection and acceptance of men for service—remembering, too, that only one rule exists for all parts of the regular army alike—a far higher standard in respect to visual power may well be looked for in the British as compared with continental armies. In a country in which *conscription* is in force, and very large armies are maintained, it is an object not to allow any men to escape conscription who can be turned to useful account in military service—if not fit for one branch, then to utilise them for some other—so that only an extreme degree of M. is allowed to exclude altogether from conscription. In a country in which *voluntary enlistment* and high wages are the rule, it is the object, as far as practicable, not to accept any men who are not sufficiently qualified for the performance of the duties which will devolve on them in the army. It is not to be forgotten that in many of the most important duties which soldiers have to perform, as sentries, videttes, in reconnoitring, in aiming at long ranges, clear vision of distant objects is essential for a trustworthy discharge of them. And when only one standard exists for all alike, it appears evident that the standard for those parts of the army in which a high degree of V. is a necessity should be chiefly taken into account in framing rules on the subject, in order that the required degree of military efficiency may be attained. To admit recruits with M. of about 7 D, with which conscripts are admitted in Italy and France, and other countries, or even with 3 D, and then to draft them for service as riflemen, would be a wasteful pecuniary outlay in the English army. Experience proves that the higher degrees of M.—M. of 7 D and above—rarely exist without the existence of posterior staphyloma and a tendency for the M. to increase. Considering, moreover, the amount of military service which is passed by a large proportion of the British army in India, and the ill effects resulting from the over-stimulation of the retina of a myope by tropical light, the expediency of admitting men into the ranks of the army with such a degree of M. as is compatible with the present test of counting the test-dots at a distance of 10 feet, appears to be very doubtful. It seems to be very questionable whether any man with M. = 1·75 D ought to be accepted as a recruit, and no doubt it would not be done if it could be avoided; such a man certainly cannot be a desirable recruit for the ranks in which the Martini-Henry rifle constitutes the firearm in ordinary use, or in any part of the service where precision of vision for distant objects is demanded unassisted by correcting spectacles. For parts of the army in which accurate sight is not such a necessity, it would be difficult to lay down any rule as to the limits of M., or other ametropic conditions, admissible; for just as the regulations regarding the height, girth of chest, and other physical conditions required in recruits are varied according as they are urgently wanted or not, or according as the supply of them is scarce or plentiful, so also it may be expected that the regulations regarding ametropia will be varied. In case there are more recruits to be had than are wanted, better qualities of V. may be insisted upon; in case the need is greater than the supply, inferior qualities of V. will have to be accepted. But obviously, when the standard for V. has to be reduced, the reduction should be confined as far as practicable to those parts of the army in which it will least interfere with the performance of the duties appertaining to them.

35. **Character of V. possessed by a Myope of 1·75 D.**—A man in early life affected with M. of about 1·75 D, sees all objects beyond two or three feet with more or less indistinctness. At the distance of 20 feet, books on shelves, or other objects of like sizes, appear mixed up together owing to want of definition and to reduplication of outlines, while, though at this distance the figure of a man may be seen well enough, his features are not separately distinguishable. An acquaintance even is not recognised at this distance if the recognition depend upon peculiarity of feature, unless the light is very strong and happens to fall directly upon the face, though striking contrasts in uniform, such as stripes on the sleeve, medals, differences in colour, or peculiarities of carriage and of movements of the body, are sufficiently obvious, even under moderate light. At a distance of 50 yards and upwards groups of five or six persons standing together before a moderately dark background cannot be readily separated from one another so as to admit of being counted with accuracy. Dark objects on a white ground, such as large black letters on a white notice-board, appear lighter, and the white ground appears darker, than they really are, while the letters are so spread out that they are altogether indistinguishable at a distance at which an eye with normal vision can recognise the painted words without difficulty. More distant objects, such as the general features of a landscape, houses and persons among trees, are huddled together and converted into little else than general shadows with intermediate outlines. The nature of particular objects, even objects of large size, can only be made out when the accidental advantage of some sharply marked contrast is afforded, such as is presented by a ship floating on water, by a building or a tree having the skyline as a background, or when a well-known object such as a horse is in movement on a road. Even this last object ceases to be distinguishable at a distance of seven or eight hundred yards if it be passing by a dark background such as a belt of trees. The want of clearness of view

increases, and the power of recognition diminishes, in proportion as the intensity of light diminishes, so that on a day when the sun is obscured by cloud, and the light therefore comparatively dull, but not so dull as to interfere with due perception if V. be normal, the power of distinguishing objects by the myope of 1·75 D, and of higher degrees, is materially curtailed. The opening of the pupil to admit more rays of light obscures the view through greater diffusion of the peripheral rays. Still more difficult does it become for such myopes to distinguish particular objects after sundown, even when there is sufficient light for men with Emc. vision to be able readily to perceive and recognise them.

It is true that such Mc. persons have the compensating advantage of seeing minute objects near to their eyes clearly, and that they retain this power at periods of life when convex spectacles have become a matter of necessity to persons with Emc. vision; but obviously this special power is of scarcely any advantage so far as military service in general is concerned, while the morbid changes to which all Mc. eyes under certain conditions are liable, especially in tropical countries, and the risk of the M. becoming progressive, cannot be ignored.

The visual difficulties just now described have to be encountered by myopes when the atmosphere is of its ordinary clearness, and when only the intensity of light is diminished. Such difficulties are greatly aggravated when the atmosphere from any cause is not clear, as when there is fog, mist, or when rain or snow is falling. Under these conditions, the view of objects becomes more or less obscured, according to the amount of rain or watery vapour in the air, to all persons, however acute their power of sight may be; but in the case of the myope the obscurity is much augmented relatively to an emmetropic person from the effects of ray-diffusion. The increase of obscurity is in proportion to the increase in the degree of M. If two persons, one Mc., the other Emc., are looking at a misty landscape, the former sees it as if he were looking through a thicker veil than the latter. When the mist is only moderate in amount, permitting all the principal objects in a landscape embracing several miles to be distinguished by the emmetrope although the foreground, middle distance, and distant hills may appear clouded and have a general grey hue, the myope finds a difficulty in recognising many of the objects before him. The whole prospect appears lighter in colour, and the principal objects in it have their outlines less defined. This happens although the degree of M. may not exceed one dioptric. The surface of the hull of a ship at anchor, which appears dark to an emmetrope, is so mixed up with the water on which it floats, and the surface of the water with that of the ship, that both assume a nearer approach to uniformity of tint, and blend together more intimately. If the myope have a higher degree of M.—a degree equivalent to 2 D, for example—the most distant objects of the landscape disappear from his view altogether, while nearer objects are rendered more indistinct. The outlines of such an object as the ship just now mentioned are still more diffused, while perhaps the mast and rigging fade from sight altogether. If the M. be still higher in degree, such as 4 D, even large objects in the foreground are quite confused; while such objects as the ship and the water on which it floats, notwithstanding the strong difference in colour which really exists between them, become indistinguishable, are not only not separately recognised, but are not visible. They are mixed up with the general haze. The retinal images of the mist and rain, like the images of other objects, are diffused and mixed together, from the scattering of the rays attendant on the Mc. formation of the eyes, and increase the general obscurity of the landscape presented to view. Even suitable lenses, that would correct the Mc. vision under most other circumstances, fail to rectify the faulty vision in such weather, for the fog or mist clouds the glass, or the rain wets it and for the time destroys its translucency.

The injurious influence on eyesight just described of such a relatively moderate degree of M. as about 1·75 D sufficiently indicates the objection to putting Mc. officers or men, especially in the dusk of twilight or when the atmosphere is misty, on any duties, the proper fulfilment of which depends either on an extensive range of clear vision and observation, or on keeping a sharp outlook with respect to objects which are neither very conspicuous from size or contrast of colour.

When the quantity of light is much lessened, as in twilight or dusk, the pupil of the Mc. eye, in common with the pupils of all eyes, dilates in order to admit more of the rays of light proceeding from objects, and in this way to make up for their diminished illumination. The vision of the emmetrope is improved by these means, notwithstanding the general obscurity, for the retinal impressions are amplified owing to the increased quantity of light admitted through the enlarged pupil. The vision of the myope, on the other hand, is on the whole rendered worse by the occurrence. A greater amount of light is equally admitted into the Mc. eye, but the dilatation of the pupil at the same time unavoidably gives access to more peripheral rays. The retinal images in consequence are obscured by a greater number of circles of diffusion, these circles are more widely spread, and increased indistinctness of view is the result.

36. **Degrees of Myopia admissible when the Test-dots are held at 15 feet.**—A myope with M. = 1·25 D still sees objects mistily, but the difference between the amount of their haziness in such a case, and in that of a myope with M. = 1·75 D, is considerable. If the test-dots were ordered to be counted, e.g. at 15 feet, myopes with M. = 1·25 D would be just capable of admission, and of

course all lower degrees of M. could pass the test; but myopes with M. = 1·75 D and upwards would be excluded.

A myopic eye of 1·25 D can read 0·5 Sn. at a distance of 20 inches, but not D = 1 Sn. at the full distance of a metre ; can recognise D = 6 Sn. in good light at about 3 metres, showing $V = \frac{1}{2}$; and has a distant point of distinct vision for 0·5 Sn. at $7\frac{1}{2}''$ with the 10″ convex lens before it.

37. **Degrees of Ametropia which exclude from Service in Continental Armies.**—In continental armies the degrees of ametropia which render men unfit for military service are always very high degrees. But it must be remembered that these armies are raised by conscription, and the degrees of ametropia which have been fixed for excluding from military service are those which are understood to cause total unfitness for military avocations, not merely such as unfit men for becoming reliable and good riflemen. The important point is not to be forgotten, too, that in armies raised by conscription, if a conscript is affected with a less degree of ametropia than the absolute disqualifying degree, but still one that unfits him for becoming a good rifleman, cavalry vidette, or artillerist, he is simply drafted to some other part of the army in which his amount of ametropia will not interfere with the right performance of the duties belonging to it. ·In some of them, too, correcting glasses are permitted to be used. The system of the British army, as already mentioned, does not arrange for distribution of men on the same principles.

The limit to which V. of the Rt. eye may be reduced without incapacitating men for military service varies in different continental armies. In the Austrian army reduction of V. of Rt. eye to $\frac{1}{2}$, in the Italian and Belgian armies to $\frac{1}{3}$, and in the French and German armies to $\frac{1}{4}$ of normal V. unfits for service in the ranks. In the Swiss army no reduction of V. of Rt. eye is permitted in the artillery, while the minimum for infantry is $\frac{2}{3}$, and for other troops $\frac{1}{2}$ of normal V. The extreme is, therefore, the limit of $\frac{1}{4}$ in the French and German armies. Supposing the average width of the body of a soldier equipped for field service to be about 18 inches, he would in this direction subtend a visual angle of 1′ at 1,710 yards' distance; and, circumstances of ground, atmosphere, light, &c., being favourable, he should be discernible by a soldier having normal V. at that distance. If V. be reduced to $\frac{1}{4}$th, he would, under like circumstances, not be seen beyond a distance of about 427 yards, and it is presumed that this distance has been regarded, in the armies before named, as sufficient for the ordinary needs of military service.

38. **Shortsighted Men at Musketry Instruction.**—One of the regulations regarding musketry instruction provides that, ' when a medical officer certifies that a man cannot see beyond a certain distance, he must fire in both range and field practices according to

his power of vision. Spectacles may be worn ' (Par. 20, p. 14, Musk. Regs. 1887). Shortsighted men being thus permitted to wear spectacles, if no ocular defect excepting shortsightedness exists and the spectacles are accurately adjusted, they will be nearly in the same condition as other persons who are not shortsighted. Other conditions being normal, the power of sight of a myope, when corrected, will be the same up to the farthest distance as that of an emmetropic man of corresponding age.

39. **Use of Spectacles by Soldiers in the Ranks not sanctioned.** —The use of spectacles by soldiers in the ranks, whether serving at home or on foreign stations, for the correction of ametropic defects of vision, is not sanctioned by any published regulation in the British army. Difficulties would be experienced if, without special arrangements, glasses were allowed to be worn in the English army as they are by parts of the troops of some continental armies which are raised by conscription. There would be the want of means of replacing them when injured or broken in many of the distant stations in which English troops are habitually employed, excepting after great delay; while, if the glasses were damaged or broken when the men were on active service out of Europe, there would frequently be almost insuperable difficulties in replacing them, and this would just be the occasion when they would be most urgently and speedily required. The want of education in respect to the use of such appliances, and the habits of a large number of the men composing the ranks of the English army have, no doubt, also been taken into account in withholding official permission for the use of spectacles by soldiers. Moreover, though the spectacles may be perfect in correction, when dust, rain, condensed moisture, or other matters interfere with their transparency, and, therefore, with their utility, accoutred infantry soldiers carrying rifles, or mounted troopers, are too fettered to be in a position to remove these impediments as they occur, and, during the time they are thus obscured, the spectacles cease to be of any advantage. These have probably been the principal reasons which have prevented the use of spectacles from being sanctioned among men in the ranks of the British army.

If the use of spectacles by men in the ranks should hereafter be sanctioned, arrangements will at the same time have to be made for enabling the men who wear them to replace them whenever they may happen to be lost or broken. Machine-made spectacles for correcting ordinary degrees of M. and H. are now manufactured so cheaply, and in many instances answer their intended purposes so fully, that, with proper examination and selection, the cost alone would hardly be an impediment to their introduction among soldiers. The great difficulty to be overcome in their introduction would be the arrangements necessary for ensuring suitable glasses being always available on occasions of need in some of the remote places where English soldiers are liable to serve.

There do not appear to be any military disadvantages or difficulties met with from the use of spectacles among troops on active service in Europe, so long as sufficient supplies are carried with the army to replace them in case of loss or accidental injury. No mention of any ill results from spectacles being worn by German soldiers is made in the official histories of the war of 1870–71 ; and I have been informed by medical officers holding high positions in the German army, that they are not aware of any trouble or inconvenience having been caused to the service, or to the officers or men themselves, by their use during the war. The employment of spectacles by soldiers in the ranks of the French army, and their gratuitous issue under certain restrictions, have been sanctioned since the war of 1870–71. (See Appendix.)

40. **Range and Power of Vision necessary for Commissioned Officers in the Army.**—The necessity for distinct vision appears to be even more imperative in officers than in private soldiers. They must look down whole lines of men, and see them with distinctness. They are required to observe distant objects and not unfrequently to give important directions according to the judgments they form of them. It is also of personal importance to combatant officers themselves, on taking the field, that they should possess a normal range and power of vision, or at least a range not far short of normal range. An officer of the regiment in which I was serving during the Crimean war declared to the commanding officer that he was unable to take picket duty, because he could not distinguish an enemy from a friend at a short distance from him, particularly in twilight. Eventually the matter was referred to a committee of medical officers, who reported the officer unfit for service in the field in consequence of his high degree of short-sightedness, and he was sent back to England. During the same campaign it certainly once happened, if not more often, that an officer was taken prisoner by walking into the midst of a party of the enemy whom he failed to distinguish from his own troops owing to the same cause. I on one occasion made a voyage with an officer who was taken prisoner under the circumstances mentioned, and sent to Russia, and the high degree of his M. was sufficiently obvious. An unfortunate English regimental surgeon in the Crimea who was affected with M. failed to recognise in twilight a French sentry from a distance in front of the lines at which the sentry saw him but too plainly, and he lost his life in consequence.

Even if eye-glasses are used in the field, besides the objections attending their brittleness of substance, their frequent displacement, the many things which from time to time are apt to interfere with their transparency, and the difficulty of finding substitutes for them when they are missing or lost on active service, have all to be taken into account.

A short-sighted officer commanding a regiment at Souakim in

1885 was temporarily disabled for duty by accidentally breaking the only eye-glass he had with him. He depended upon it for a clear view of objects. There were no means of replacing it by purchase nearer than Cairo; but fortunately a medical officer, also Mc., though less so, was found who was able to spare one of his duplicates, and thus materially to lessen the inconvenience to which the commanding officer had become subjected.

The authority under which a few British officers have worn spectacles in the field appears to have been one of special permission. I am not aware that any regulation sanctioning their use by officers on active service has ever been published. An officer of high position, whose son had failed to pass the regulated tests of sight, remarked to me that the same regulations would have prevented the admission into the army of Genl. Sir Charles Napier. It certainly would have done so, for I have had the opportunity of examining the spectacles successively worn by that most distinguished general officer, and I found that his M., which in his early days was as high as 7 D, being progressive, became gradually increased to 14 D. But the difficulties Sir Charles Napier had to contend against in consequence of his extreme short-sightedness can hardly be estimated, while with the eminent abilities and strong will which distinguished him he would certainly have been at least equally great as a ruler and commander had he been free from visual defect. He himself has recorded that, at the battle of Corunna, being without spectacles and finding himself alone at one part of the action, he 'felt great fear, for his short-sightedness disabled him from seeing what was going on, and what was to be met;' and subsequently, referring to this occurrence, he remarked, 'when wearing spectacles, the nervous feeling is not so strong, but the disadvantage of bad sight is tremendous when alone, and gives a feeling of helplessness.'[1]

41. **Loss of Sight of one Eye.**—Loss of sight of one eye from any cause, however good the sight of the other eye, totally disqualifies for an army commission. It is obvious that if a candidate were passed fit for a commission, notwithstanding loss of sight in one eye, he would not only be less efficient from loss of binocular V., but any injury interfering with or destroying sight in the remaining eye after he had entered the service would entirely disable him from further duty. Other considerations arise when the sight of one eye of an officer already serving in the army is lost, while the other remains effective; and consequently officers are occasionally to be met with, as well as men in the ranks, who only have the sight of one eye left. But their efficiency in various

---

[1] *Life of General Sir C. Napier*, 1857, i. 101, 111. In a voluntary service, such as the British service is, it must be remembered that every short-sighted cadet admitted takes the place of a cadet with normal sight; while in regard to knowledge and abilities, so far as the evidence of the examinations for admission allow comparison, both cadets may be on an equal footing.

respects is nevertheless considerably lessened.   (See Binocular and Monocular Vision, p. 3.)

42. Declaration as to Sight by Candidates for Army Commissions.—It has hitherto been the custom for candidates for commissions in the army to sign a declaration concerning their general state of health prior to undergoing physical examination, especially that 'vision is good with either eye (with or without the aid of glasses, as the case may be).' Remembering, however, that persons have very different views as to what constitutes 'good vision,' many knowing no other standard of quality than that of their own sight, it is obvious that an independent examination to ascertain the real quality of V. in each instance, or, at any rate, to determine that it is not below some definite authorised standard, is far more reliable and useful than any such general statement can be.   The declaration had the advantage of being likely to prevent a candidate submitting himself to examination who had any serious defect of vision in either eye.

43. Visual Examination of Candidates for Commissions in the Line.—The rule lately in force for the visual examination of candidates for commissions in the line was as follows : Simple M. or H. was not held to be a disqualifying condition.   So long as these defects could be corrected by suitable glasses they were not to exclude from admission.   The practice was, if the candidates could count the test-dots at the ordinary distances, with or without the aid of glasses, they were accepted for service so far as vision was concerned.   The test-dots were used in the same manner as with men seeking enlistment in the ranks.[1]   No rule was laid down regarding the manner in which candidates for commissions were to employ glasses in the correction of visual defects.   The responsibility regarding this point rested with the examining medical officer. In some instances candidates were allowed to use the glasses with

---

[1] *Deception Practised at Visual Examinations* —Attempts to impose upon examining surgeons may occasionally meet with success. Some years ago I was asked by the friends of a youth who was anxious to obtain a military career, but whose sight was defective in one eye, whether any likelihood existed of his being able to pass the physical examination for a commission. I found on examination that although V. was normal in the right eve, only quantitative V. existed in the left eye. I therefore said there was not the least chance of his passing the examination, as each eye would be subjected to tests separately. The next time I met him he was going through the course of study at Sandhurst. I inquired how le had contrived to get through the examination as to power of sight, and he at once showed me his plan of proceeding. When he was asked to count the test dots with his right eye he did so easily, at the same time covering his left eye with his *left* hand. He was then told to count them with the other eye. In answer to this direction he covered the left eye, the same he had covered before, but this time with his *right* hand, and of course, as before, counted the dots without difficulty. He had changed hands but not eyes, and this he did so quickly and smartly, that the examiner, who may at the moment have been observing the test-dots to ascertain if the right number were stated, did not notice the feat of legerdemain which was being practised on him, and passed him as optically fit for service. I am informed that some candidates recently have purchased Snellen's sets of metrical test-types and learned the series of letters and printed paragraphs off by heart.

which they had provided themselves prior to the examination ; in others they were required to select the glasses which suited them from a number of pairs of spectacles placed on a table in the examination room.

In March, 1887, the instructions for the examination were changed. Snellen's metrical test-types were ordered to be substituted for the test-dots. The candidate was 'to be able to read ordinary type, distinguish the principal colours, and to perform routine duties without glasses,' but for other purposes the use of glasses was permitted. If, with the aid of glasses, 'the range and quality of V. were rendered normal, and if there were no evidence of organic disease or of progressive deterioration of V., and the candidate were in other respects eligible, he might be accepted.' It will be noticed that these regulations left the decision as to the size of the type called 'ordinary type' to the judgment of the examiner, and also the particular nature of the duties called 'routine duties' which the candidate had to perform without glasses. Practitioners in civil life, when consulted by parents who were anxious to ascertain the visual fitness of their sons for military service before putting them to special courses of education to enable them to compete for commissions, experienced difficulty in expressing a decided opinion on the subject, and their representations led eventually to the further alterations which are described in the next paragraph.

44. **Power of Sight now required in Candidates for Commissions in the Army.**—The latest rules for the visual examination of candidates for army commissions, as directed in an army circular dated September 1887, are simple and precise. They may be summarised as follows:

1. If the candidate can read the small Sn. type $D = 0.6$ at any distance, and at the same time can read Sn. $D = 6$ at its normal distance, without glasses, he is to be passed as visually fit.

2. If the candidate cannot read with each eye, without glasses, Sn. large type, such as $D = 36$, at one-sixth of its normal distance, he is unfit.

3. If the candidate can just read with each eye, without glasses, Sn. large type at one-sixth of its normal distance, but not beyond it owing to faulty refraction, he may be passed fit, provided the distant V. of one eye can be fully corrected by the aid of glasses, and the distant V. of the other eye can be corrected up to not less than one-half of normal V., and provided also that the candidate can accommodate for the type $D = 0.8$ at any near distance.

4. Strabismus, achromotopsia, and any morbid conditions subject to the risk of aggravation or recurrence in either eye, will cause the rejection of a candidate.

The regulations are given *in extenso* in the Appendix (See p. 225).

45. Indian Government Service.—Candidates for admission into the several departments of H.M.'s Indian Government Service are ordered to be tested as to V. by letters corresponding to those in the last edition (1885) of Snellen's test-types. These test-types, as in all the later editions, are numbered on the metrical system. The person to be examined is to be placed with his back to the light at a distance of 6 metres from the types, which are to be hung or held perfectly upright with a bright light falling on them. The ophthalmoscope and retinoscopy are to be employed when necessary. It is further directed that the existence and degree of ametropia are to be carefully recorded in all cases of examination.

The visual regulations affecting special parts of the Indian Government Service are given under their respective headings. (See also Appendix, p. 226.)

46. Visual Qualification for the Indian Government Civil Service.—A candidate for this branch of the public service is not disqualified by ametropia of one or both eyes, provided V. can be brought up to $\frac{6}{9}$ in one eye, and to $\frac{6}{9}$ in the other eye by correcting lenses, and that there are no morbid changes in the fundus of either eye. A posterior staphyloma, however, if stationary, will not disqualify a candidate, provided the accompanying M. do not exceed 2·5 D in either eye, and no active morbid changes of the choroid or retina be present.

Nebula of the cornea of either eye disqualifies if V. be reduced to less than $\frac{6}{12}$; and in any case where nebula exists, V. in the better eye should equal $\frac{6}{9}$ with or without glasses.

Paralysis of one or more of the exterior muscles of the eyeball disqualifies a candidate, but a candidate who is said to have been cured from squint by operation may be accepted, although binocular vision has not been restored, provided V. can be brought up to $\frac{6}{9}$ in one eye, and to $\frac{6}{9}$ in the other eye, by correcting glasses, and the movements of each eye be good.

The foregoing rules also apply to the Chaplains' and Educational Departments of the Indian Service.

47. Visual Qualification for the Public Works Department under the Indian Government.—Candidates for admission into this branch of the Indian Service are subject to the following rules :

A candidate may be accepted if, being Mc. in one or both eyes, the M. do not exceed 2·5 D, and if, with correcting glasses, not exceeding 2·5 D, V. in one eye is $= \frac{6}{6}$, and in the other is not less than $\frac{6}{9}$, while the range of Acc. is normal. The same rule applies to Mc. astigmatism. There must be no evidence of progressive disease in the choroid or retina.

A candidate having H. not exceeding 4 D is not disqualified, provided, when under the influence of atropine, V. in one eye is $= \frac{6}{9}$ and not less than $\frac{6}{9}$ in the other, with + 4 D or correcting lenses of any lower power. The rule is similar as regards Hc. astigmatism.

Corneal nebula disqualifies if V. in one eye is less than $\frac{6}{12}$. If V. in one eye is not below $\frac{6}{12}$ when corneal nebula exists, V. in the other eye must be normal, and the eye be Emc. for the candidate to be accepted. Defects of V. from pathological changes in the deeper structures of either eye may exclude a candidate from entry into the service.

Inability to distinguish the principal colours (Achromatopsia) is a disqualifying defect; as is also paralysis of one or more of the external muscles of the eyeball.

Admission into certain other departments of the Indian Government Service, as the Telegraph, Forest, Survey, Railway, Factories, and Police departments, is regulated by the same visual limitations as those above given.

48. **Visual Qualification for the Army Medical Staff.**—In the 'Schedule of Qualifications necessary for Candidates desirous of obtaining Commissions in the Army Medical Staff,' it is ordered that a Board of Medical Officers must certify the candidate's 'vision is sufficiently good to enable him to perform any surgical operation without the aid of glasses. A moderate degree of M. will not be considered a disqualification, provided it does not necessitate the use of glasses during the performance of operations, and that no organic disease of the eyes exists.'

The objection to dependence on the use of glasses for securing clear vision in the performance of surgical operations is a well-founded one. Irrespective of the inconvenience which may occasionally arise from glasses falling off in the movements of the operator at critical moments, they are constantly exposed to loss of transparency from condensed vapour upon them, or from becoming spotted by jets of blood.

So far as concerns the refractive defect of M. mentioned in the schedule, it will not be difficult for a moderately Mc. candidate to comply with the condition named, for the M. must be of very high degree to necessitate the use of glasses in performing most surgical operations. With M. of 2·0 D, a candidate of twenty-five years of age, possessing normal power of accommodation, would have a range of clear vision for small objects from a distance of 20 inches to a point a little within 4 inches, without the aid of glasses, and, indeed, within the range named would see more comfortably without glasses than with them, though his vision for more distant objects would be very imperfect unless helped by suitable glasses. Even with M. of 2·50 D he would have a range of clear vision from a distance of 16 inches to a little over $3\frac{1}{4}$ inches. But with H. of 2·50 D, when, at the age named, only about six dioptrics of accommodation would be available for work at near objects, even with the highest accommodatory effort a near point of distinct vision could only be obtained at a little under 7 inches. Short of that distance vision by the unaided eye must be more or less obscured; and since it is not possible to continue to exercise Acc. at its highest state of tension, and only about half of the 8·50 D of the Acc. belonging to the age named could be continuously employed, the nearest point of distinctness for prolonged vision would be removed to 20 inches' distance. Under such circumstances attempts to distinguish minute structures of importance, probably confused by blood, to separate a small artery from an adjoining nerve, for example, without the aid of suitable glasses, would be very embarrassing, and might lead to serious errors. It is therefore the Hc. eye which would meet with the most difficulty in operating without glasses, and this difficulty would increase with increase of age. If the H. were complicated by Ast., the difficulties of the operator would be considerably increased.

49. Visual Qualification for the Indian Medical Service.—Candidates for admission into this branch of H.M.'s Service may be accepted if, being Mc., the M. do not exceed 5 D, and if, with correcting lenses, V. is brought up to $\frac{6}{6}$ in one eye, and to a standard not below $\frac{6}{12}$ in the other eye. If the M. be attended with posterior staphyloma, the M. must not exceed 2·5 D in either eye; and V. must be capable of being brought by correcting glasses up to $\frac{6}{6}$ in one eye and not below $\frac{6}{12}$ in the other. There must be no active morbid changes in the choroid or retina.

If the M. be accompanied with Ast., a candidate will not be disqualified if the combined spherical and cylindrical glasses for their correction do not exceed −5 D, and if, when thus corrected, V. in the two eyes is brought to the standards named above and Acc. is found to be normal. No progressive morbid changes of the choroid or retina must be present.

If the candidate be Hc., he may be accepted provided his H. do not exceed 5 D, and, when the eyes are under the influence of atropine, V. in one eye is brought to $\frac{6}{6}$ and in the other to at least $\frac{6}{12}$ with the aid of +5 D or lenses of any lower power.

If the H. be complicated with Ast., the candidate may be accepted provided the combined correcting lenses do not exceed 5 D, and V. can be brought to the standard of $\frac{6}{6}$ in one eye and not less than $\frac{6}{12}$ in the other eye.

A candidate must be able to distinguish the principal colours, red, green, violet, blue, yellow, and their various shades.

A faint corneal nebula reducing V. in the nebulous eye to $\frac{6}{12}$ will not exclude a candidate, if the eye be otherwise healthy, and if the other eye be Emc. and its V. is normal.

A candidate will be disqualified if he suffers from paralysis of one or more of the external muscles of the eyeball.

50. Visual Acuteness required in Candidates for Commissions in the Royal Artillery and Royal Engineers.—The V. of candidates for admission into the Royal Military Academy at Woolwich until recently was tested under a special regulation. They now fall

under the same general rules as are applied to candidates for commissions in other parts of the army. These rules have been stated
in Par. 44.

**51. Visual Acuteness required in Candidates for Commissions in
the Royal Navy.**—Candidates for commissions in the Royal Navy
are not considered eligible who are subjects of any degrees of M.
or H. They are required to have visual power sufficient to enable
them to see Snellen's test-types at their full distances. The conditions of service afloat prevent naval officers and seamen from wearing spectacles, so that the correction of refractive defects among
them by these means, whenever and to whatever degree they may
exist, is impracticable. Exceptions are, however, made in some
special classes of officers, but the discretionary margin allowed is
very limited. It would manifestly be unnecessary to apply the
same visual tests as rigidly to naval chaplains and clerks as to
navigating officers.

**52. Visual Qualifications for the Marine Service of the Indian
Government.**—A candidate, in order to qualify for this service, must
not have any refractive defect of either eye which is not neutralised
by a + or − 1·0 D lens, or by a lens of some lower power; must
be capable of distinguishing the prismatic colours; and must be
free from squint or any defective action of the exterior muscles of
the eyeball. For the Pilot service candidates must have both eyes
Emc., with normal V. and Acc., and must possess perfect colour
sense.

**53. Quality of Sight for Special Course of Instruction in the
School of Musketry.**—Officers and non-commissioned officers who
desire to attend the special rifle instruction given at the School of
Musketry are not now prevented from attending in consequence
of any special degree of defective eyesight. Major Salmond,
Deputy-Assistant Adjutant-General at the Hythe School of
Musketry, has kindly informed me that the following is the regulation now in force on the subject: 'Officers and non-commissioned
officers who are detailed to attend a course of training at the
School of Musketry must (except in the case of Adjutants of
Auxiliary Forces) be examined by a medical officer as to the state
of their eyesight, which should be tested in the usual way with
test-dots, and reported on accordingly in the column for the purpose
in the printed War Office form of Return No. $\frac{104}{\text{Parties}}$, or, when
that form is not furnished, in a manuscript report. In testing
vision under this paragraph the officer or sergeant may be allowed
to use an eye-glass or spectacles.' It is required that the condition
of eyesight should be known, so that, should a physical defect of the
kind cause the subject of it to fail in attaining a certain standard
in 'range practice,' the case may be considered and dealt with by
the proper authorities.

54. **Disease of some of the posterior parts of the Eye.**—When the ocular examination of a recruit or soldier leads to a suspicion that he labours under impaired vision owing to deeply-seated disease of the eye, there is only one way of determining the correctness or incorrectness of the suspicion, and that is by ophthalmoscopic examination. A description of the method of employing the ophthalmoscope, as well as of the other steps to be taken, for establishing a correct diagnosis of the different morbid states of the posterior parts of the eye, is to be found in all systematic works on Diseases of the Eye.

55. **Impaired Vision, or Blindness, of one Eye in a Trained Soldier.**—It has been mentioned with regard to a recruit that any important visual defect in either eye renders the man unfit to engage for service as a soldier. The rule is different with regard to men who are already serving in the ranks. Impaired visual power, or total loss of vision of one eye, if the other eye be efficient, is not held to be a cause of unfitness for further military service. This rule is laid down in War Office Circular, No. 874, of the 17th August, 1864, in the following terms: 'No soldier shall be discharged for the loss of one eye only, whether it be the right or the left; but if a soldier shall have lost one eye by a wound in action, or by the effects of service, and shall receive other wounds or injuries in action, or be otherwise so disabled as to render his discharge necessary, the loss of an eye shall be taken into consideration in fixing the pension at such a rate as his combined wounds or disabilities may entitle him to receive.'

56. **Aiming with the Left Eye.**—If a soldier, after enlistment, is found to have a defect of vision of the right eye, which has not been previously detected, and it is found to incapacitate him for using his rifle from his right shoulder, or if the right eye of a soldier becomes disabled by disease or injury, he is permitted, under certain rules, to fire from the left shoulder—thus using his left eye for sighting the objects aimed at.

This permission is never granted excepting under certificate from a medical officer that the soldier is labouring under defective vision of the right eye. Now that skirmishing and independent firing are so much more employed than firing in close order, it is of less consequence that a man's mode of firing differs in the respect named from that of the other men of his company. Moreover, the permission to aim with the left eye is only turned to practical account at target practice, or when firing with ball cartridge.

57. **Assumed Blindness of one Eye.**—Blindness of the right eye --hemi-anopsia—is not unfrequently simulated in foreign armies to escape conscription, but English soldiers very rarely make pretence of blindness of one eye, because it is well known among them that uniocular loss of V. does not incapacitate for further military service. Should, however, blindness of one eye be alleged to exist,

and no objective signs to warrant the assertion be obvious, so that a suspicion of simulation be excited, there are various tests which may be resorted to for determining whether the suspicion is well grounded or not.

In the first place, the observation of the surgeon may be directed to the indications afforded by the action of the iris belonging to the supposed blind eye. In complete blindness of one eye, the pupil is partially dilated owing to the absence of all reflex stimulus from the insensible retina. If, then, the sound eye is excluded from light, and the iris of the alleged blind eye is found not to contract when it has been shaded by the hand of the surgeon and then suddenly subjected to the admission of strong light, but is found to contract when the retina of the other eye is exposed to sudden access of light under similar conditions, it is evident that the alleged existence of blindness is real; the iris is not stimulated to contract by the action of light on the retina with which it is directly connected, but contracts under the influence of light on the other retina with which it is indirectly associated; on the other hand, if the iris of the alleged blind eye does contract when its retina is exposed to sudden light, while the other eye is kept closed, there may be a considerable amount of amblyopia, but there cannot be complete amaurosis.

If the iris of the alleged blind eye do not answer to the stimulus of light, either when its own retina or that of the other eye is excited, the alleged blindness may or may not exist; the observation only proves that there is paralysis of the iris, and this may be the result either of natural causes or may be artificially induced by the application of atropine, henbane, daturine, or other mydriatics. For determining the real cause of the paralysis, other investigations must be undertaken, but these need not be described here.

The effect of covering the alleged blind eye, and then of suddenly exposing it to light, on the other, or acknowledged seeing eye, may also be observed. When the two eyes are in a normal condition, the effect of shading one of them from light is to produce a slight increase in the size of the pupil of the other eye; on removing the shade and exposing the eye suddenly to light, a slight decrease in the size of the pupil of the eye that has not been shaded may be noticed. If the retina of one eye be insensible to light, neither shading nor uncovering this blind eye will produce any change in the dimensions of the pupil of the other eye. The seeing eye during the experiment should be directed to look at some distant object so that neither Acc. nor convergence may be exerted.

But the most effective means for unmasking an attempt at deception of this sort among comparatively uneducated persons is Græfe's prism test. If a prism of 12° or more be held with its base upwards or downwards before the eye in which visual power is acknowledged to be retained, and the person who is subjected to the test, on being asked what effect it has on his sight, states that the glass in front of his eye causes him to see double, the simulation is proved, for diplopia could only result by both eyes seeing. If the person under examination is educated, and a page of print is placed before him, while the prism is placed in front of his seeing eye with its base upwards, and he is then unable to read without great difficulty, it is another proof that he is seeing with both eyes, and that his reading power is interfered with from the images of the print overlapping; if he really had only monocular vision, he would see the print as well with the prism as he did without it, though it would be a little altered in its apparent position. Again, if the base of the prism be turned horizontally inwards, and the eyes then squint, it is proved that an effort is being made to prevent double vision, and that, therefore, the assertion of blindness of one eye is untrue. By getting the person to read some of Snellen's test-types, and to describe first one of the two images and then the other, and by varying the sizes of the objects presented, the surgeon may arrive at a conclusion as to whether any amblyopia exists or not in the alleged blind eye, and, if it do exist, may even a certain its degree.

Example.—The following case, taken from the records at Netley, will serve to illustrate the application of this test. Private T. F., 84th Regiment, was invalided in 1866 from Malta, and admitted at Netley, under Amaurosis. His condition at Netley he stated to be: right eye quite blind, no perception of light; left eye, reads No. 4 Snellen at 3″, counts fingers at 3′. No ocular abnormality was visible by ordinary or under ophthalmoscopic examination. The left eye was kept bandaged for a couple of days, on the plea of resting the eye, but really to observe whether he could guide his movements by the right (alleged totally blind) eye. While thus bandaged he was reported to have been seen reading, or apparently reading, a book. I then tried the prism test, and the man described two images of a single object, a lead pencil, held at a distance of about 4′ from him, together with the movement of one round the other as the prism was made to revolve before his left (acknowledged seeing) eye. The imposture being thus proved, the man was discharged to duty at his depôt.

The stereoscope has been used for detecting simulated blindness of one eye, and is still more puzzling to one who is not acquainted with its effects. The particulars of a case were related to me in which imposture by a foreign officer who simulated blindness of one eye as a result of field service in the war of 1870–71 was fully detected by its means, and in this instance the use of a simple prism, owing to the intelligence of the person examined by it, and his knowledge of the effects of prisms, had failed to prove the deception. It is hardly possible for a person who sees with both eyes to answer the stereoscopic tests in the same manner as if one eye were blind, provided the experiment is fairly performed. But the surgeon must be on his guard that the person under examination does not close the eye which is alleged to be blind while the objects placed in the stereoscope are exposed to his view. The eyes must be watched during the examination. The stereoscopic objects should be specially prepared. Series of lines, different in colour—red and blue, for example—so arranged that in the combined image they cross each other in direction, may be used for the purpose. When both eyes are sensible to light, the lines are seen constantly changing, more or less completely, from one colour to the other; and it is not possible to say by which eye either coloured lines are seen. A person blind of one eye will see the lines of one colour only. Two printed paragraphs, equal in size and similar in character, but differing in parts of the text, may be placed on the stereoscopic slide. A person regarding the slide with both eyes through the stereoscope will not be able to read the portions where the texts differ, for the print of one side will be mixed up with the print of the other and there will be a constant interchange of text, and struggle between the fields of vision of the two eyes, so as to make reading impracticable. A person who does not see with one of his eyes will read easily the print presented to the seeing eye. Figures in endless varieties of shapes and colours may be employed in a similar manner; so that no simulator who is capable of seeing with both his eyes the objects presented to him in quick succession on the stereoscopic slides, when describing what he sees, can help including those parts which could only be visible to him through the agency of his pretended blind eye, unless he succeeds in temporarily closing it for the purpose of excluding its view.

Staff-Surgeon Dr. Burchardt, of Berlin, invented a portable stereoscope, of very ingenious construction, for the practical diagnosis of simulation of blindness or amblyopia of one eye.[1] The test-objects supplied with the apparatus are very effective. The stereoscope is of the lenticular form, and the lens sections are so arranged that at the same time that the person under observation is looking through them at the test-objects, the examiner can look through them and obtain a magnified view of the eyes of the person he is examining. He can thus ascertain that neither of the eyes of the supposed simulator is closed even for a second during the application of the tests.

In exceptional instances, a person who is ametropic, and who has at the same time a considerable amount of anisometropia, will occasionally elude the stereoscopic test, by regarding only the side which, in respect to distance, is most in focus with one of his eyes, and suppressing the less distinct image in the other eye. In such a case the effects of complementary colours may be turned to useful account. The same test may be advantageously employed when a

[1] *Praktische Diagnostik der Simulationen von Gefühlslähmung von Schwerhörigkeit und von Schwachsichtigkeit,* herausgegeben von Dr. Max Burchardt, Obers absarzt 2. Cl. u. Privat-Docent. Berlin, 1875. Verlag der Gutmann'schen Buchhandlung. Otto Keslin.

prism or stereoscope is not at hand. A few words written on a black board with green or red chalk, or the red and green letters on a black background, or the coloured types on the grey background, in the 1885 edition of Snellen's test-types, may be used for this purpose. If the type to which the attention of the person under examination is called be green in colour and on a black background, a spectacle frame is applied with a red glass on the side of the acknowledged sound eye, and ordinary colourless glass before the supposed blind eye, and the person is asked to read the green type. If the red type is shown, a green glass is substituted for the red one in the frame. If he is able to read the type with such a pair of glasses before his eyes it is obvious that he is reading it with the alleged blind eye. As the coloured glass causes the type of the complementary colour to fade from view and to become mingled with the general black background, the acknowledged sound eye, under the conditions described, could not read them. If the type be on a grey background, its special colour will disappear, and it will seem to be dark by contrast with the surrounding grey. In this trial, the colour of the type is asked, and if in the reply its special colour is stated, it is again evident that it is with the alleged blind eye that it is seen. The coloured letters should be free from superficial lustre.

58. **Suspected Simulation of M. of high degree.**—The mode of determining the degree of M. in a given case objectively, and independently of the subject's control, has been explained in the chapter on Retinoscopy. The following example shows that, in a suspected instance of feigning, the doubt may be solved subjectively without the aid of the ophthalmoscope, but with the help of Snellen's test-types and a $+4$ D lens alone.

Example—Corporal A., æt. 21 years, was invalided to Netley for very defective vision of the right eye, due to a high degree of M. He saw fairly well with the left eye. Circumstances excited a suspicion that he was feigning, or at least exaggerating, the disability of the right eye, and he was subjected to the following tests. When Sn. types D = 6 were placed before him, the left eye being excluded from vision, he could only read them at 4 metres, about 13 feet, and then with difficulty. Sn. D = 12 being placed before him, he only read them at 8 metres, or about 26 feet. In each case, therefore, the trial showed V = $\frac{2}{3}$. With the stenopæic hole he saw both D = 6 and D = 12 much clearer, and further off.

Sn. D = 0·5 was now placed before him. He read it, but stated he could not read it farther off than 8 inches. Supposing this to be the distant point of distinct V. for this type, his M. would seem to be = 5 D. He was now tried for his near point of distinct V. with the 0·5 type, and he fixed it at 2·6 inches. This agreed with his other statements, for Acc. = 10 D (the Acc. for his age) + M = 5 D would be = 15 D, or 2·6 inches.

He was now asked to read the same type, Sn. D = 0·5, through a $+4$ D spectacle, and he read it at 4·4 inches, but not beyond. This was again consistent with his other statements, for $\frac{1}{4\cdot4}$ or $\frac{5}{22} - \frac{1}{10} = \frac{1}{8}$ nearly, or 9 D − 4 D = 5 D, confirming his alleged distant point of distinct V., viz. 8 inches without the glass.

The fact of the whole series of trials leading to consistent resu'ts left no doubt that the R. eye was Mc = 5 D, and the existence of the man's alleged disability was thus verified.

59. Assumed extreme Defective V. of both Eyes.

—Supposing a soldier maintains that he cannot see objects clearly at any range, or only within a very limited range—that no description of lens improves his vision—while no cause for the alleged disability can be discovered, ophthalmoscopically or otherwise, but, on the contrary, both eyes appear to be normal in condition, can it be proved that he really does see clearly enough for duty? To a certain extent the surgeon must be guided in such a case by circumstantial evidence and observation. A cross-examination, instituted with ordinary judgment, will usually expose the attempt at fraud. But even here the lenses can generally be turned to important use; with the convex spectacles before his eyes, a trickster will most certainly become confused, and will either see types or dots at distances which show that his vision is normal for distant objects, or he will overstate his case by giving only negative replies, saying that he cannot see at all at any distance. When this last-named position is maintained by a simulator, the surgeon can only hope to expose the deception by contriving in some unexpected way to obtain positive evidence showing that the man's statements are untrue. Many ingenious traps, some of doubtful propriety, have been devised with this object in view, and various cases have been recorded in which the simulator has been caught unawares, and the deception exposed, by their means. A harmless trap of the kind was laid successfully in the following case.

Case.—Private G. McA , æt. 19, was sent after enlistment to the depôt at Parkhurst. Subsequently to his arrival there he declared his sight to be extremely defective, and he was reported to be unable to learn his drill on this account. He was then examined by a medical board at Portsmouth. The board, not finding anything to explain the alleged condition, sent him back to Parkhurst. The surgeon in charge again after some time reported his inability to serve as a soldier on account of defective sight, and he was ordered to Netley for observations. At Netley, according to statement, 100' Snellen could not be seen beyond 2½'; fingers could not be counted farther off than 1', and hardly at that distance. Nothing abnormal could be detected in either eye by ophthalmoscopic or ordinary examination. When tried with lenses all replies were negative; he persisted in saying that he could see nothing through them at any distance. The extreme degree of amblyopia complained of was inconsistent with many of the man's observed actions, but it was difficult to get a positive proof sufficient to convince others of the deception, and a trap was therefore laid for him. While walking he was suddenly told by a sergeant in the presence of a witness to pick up a pin which had been purposely placed a little way off on the floor before him. The man, being taken unawares, at once stooped and picked it up. He was sent back to the depôt and made no further complaint of weak sight.

If the man admits that he can see objects of known dimensions, the hands and figures of a clock for example, up to some particular distance, the surgeon can notice whether, after varied changes in position of the object, he always returns to the same distance as his limit of clear vision.

If the man is able to read, the distance at which a given type of Snellen's standard is read may be noted, and observation made whether he reads the larger types, illumination and other conditions being alike, at proportionate distances. If he should state he is not able to do so, and he still maintains that neither concave nor convex glasses make distant objects clearer, and there is no evidence of the presence of astigmatism, it is obvious that he is trying to deceive. Where there is a command of lenses of many varieties of focal range, the demonstration that he is simulating his alleged defect of sight is comparatively easy.

I would not, however, wish it to be inferred from the above remarks, that I believe attempts at fraud, in respect to defective vision, are frequently to be met with among soldiers in the ranks. On the contrary, my experience leads me to believe that in the greater number of instances which have been suspected to be instances of deception, real disabilities have existed, though their true natures have not been ascertained. The patients have been supposed to be malingering because sufficient time and attention has not been devoted to the elucidation of their cases, or because the determination of the existence of the particular defective conditions of sight under which they were labouring was not included in the surgeon's range of diagnosis.

CHAPTER X.

APPENDIX OF EXTRACTS FROM MUSKETRY AND OTHER REGULATIONS CONCERNING EYESIGHT WHICH AFFECT MEDICAL OFFICERS; WITH NOTES ON CERTAIN OTHER MATTERS REFERRED TO IN THE BODY OF THE WORK.

Quality of Eyesight for Musketry Practice—Spectacles—Firing from Left Shoulder —Good Sight and Proficiency in Shooting—Aiming Drill—Judging Distance Drill—Attendance of Medical Officers at Target Practice—Range of V. of Candidates for the Royal Military Academy at Woolwich—For Commissioned Officers of the Army—Militia Regulations regarding V.—Indian Government Services—Admiralty Instructions concerning Eyesight—Rules in the German Navy—Degrees of Visual Deficiency, M., &c., which exclude from Military Service in Foreign Armies: Holland, France, Germany, Italy, Austria, Belgium, Switzerland, Denmark, Spain, and Portugal.

Quality of Eyesight for Musketry Practice.—'No soldier is on any account to be exempted from musketry training on the plea of alleged bad sight, or on any plea which, if valid, would tend to prove him unfit for service. When the medical officer certifies that a man cannot see beyond a certain distance, he must fire in both "range" and "field" practices according to his powers of vision, and his points must be included in the total from which the merit of the practice is calculated. Spectacles may be worn. If the medical officer certifies that a man has defective vision of the right eye, he may fire from the left shoulder.'—(Regs. for Musketry Instruction, 1887, par. 20, p. 14.)

Good Sight and Proficiency in Shooting.—'It cannot be too strongly impressed upon the recruit that any man who has no defect in his sight can be made a fairly good shot, and that no perfection he may have attained in other parts of his drill can, when on service, remedy any want of proficiency in shooting.'—(Regs. cit., par. 24, Ch. II. p. 17.)

Aiming Drill.—' Aim, which must be an exactly true one, must be taken along the bottom of the notch, or the top of the centre white line of the backsight, and the tip of the foresight to the centre of the mark aimed at.'

'The eye must be fixed on the mark aimed at, and not on the foresight. Recruits are apt to fix the eye on the foresight instead of the mark, in which case the mark cannot be distinctly seen, and the difficulty of aiming is greatly increased.'

' In taking aim the left eye must be closed. If a recruit is not able to do this at first, he will soon succeed by tying a handkerchief over his left eye.'

' It cannot be too strongly impressed on every man, that to shoot well at long ranges, he must train and strengthen his eye by looking at *small* objects at *long* distances. Short-sighted men should aim at distances according to the power of their eyesight; they may wear spectacles.'—(Regs. cit., pp. 76–78, paras. 54 and 57.)

Judging Distance Drill.—'The object is to teach the soldier to judge with tolerable accuracy the distance he is from the object, so that he may be able to regulate the elevation of his rifle. He will be taught to judge distance by sight and by sound.'

'Instructors must bear in mind that the eyesight of different men varies considerably, and that consequently they must not expect the observations and answers of every man will be the same.'—(Regs. cit., pp. 98–100, par. 101.)

' Between the flash or smoke of a rifle or gun and the sound of the report a certain time elapses. If this period of time be carefully noted in seconds, the distance the sound has travelled can then be calculated. Judging distance by sound can only be looked upon as auxiliary to judging by sight. It is the only method by which distance can be judged at night.'—(Regs. cit , paras. 106 and 198.)

Attendance of Medical Officers at Target Practice.—' The attendance of medical officers at target practice under ordinary circumstances is unnecessary, but the name and address of a medical officer, to whom application could be made in case of accidents, is to be communicated to all officers in charge of parties proceeding to target practice : this officer is not to be absent from his quarters or hospital during the period the target practice is being carried on. Should exceptional circumstances arise which would appear to render expedient the presence of a medical officer on a range, the general or other officer commanding may, after consulting the principal medical officer, direct the attendance of one when necessary. The principal medical officer is to report all cases of this nature for the information of the Director-General of the Army Medical Department, in order that timely provision may be made for the performance of the duties of the medical officer so employed.'—(Regs. cit., par. 124, p. 107.)

Quality of V. necessary for Commissioned Officers in the British Army.—The following Regulations and Instructions for the visual examination of candidates for commissions were promulgated to the Army by direction of the Secretary of State for War on September 1, 1887 :—

1013*b*. Letters and numbers corresponding to Snellen's metrical test-types (edition 1885) will be used for testing the standard of vision. If a candidate's vision, measured by Snellen's test-types, be such that he can read the types numbered $D = 6$ at 6 metres or 20 English feet, and the types numbered $D = 0.6$ at any distance selected by himself, with each eye separately and without glasses, he will be considered fit.

1013*c*. If a candidate cannot read with each eye separately, without glasses, Snellen's types marked $D = 36$ at a distance of 6 metres or 20 English feet, i.e. if he do not possess one-sixth of Snellen's standard of normal acuteness of vision, although he may be able to read the types $D = 0.6$ at some distance with each eye, he will be considered unfit.

1013*d*. If a candidate can read with each eye separately Snellen's types numbered $D = 36$ at a distance of 6 metres or 20 English feet without glasses, but cannot read them beyond that distance, i.e. if he just possesses one-sixth of normal acuteness of vision, and his visual deficiency is due to faulty refraction, he may be passed as fit, provided that, with the aid of correcting glasses, he can read Snellen's type $D = 6$ at 6 metres or 20 English feet, with one eye, and at least Snellen's types $D = 12$ at 6 metres or 20 English feet, with the other eye ; and, at the same time, can read Snellen's type marked $D = 0.8$ with one or both eyes, without the aid of glasses, at any distance the candidate may select.

Squint, inability to distinguish the principal colours, or any morbid condition, subject to the risk of aggravation or recurrence in either eye, will cause the rejection of a candidate.

Quality of Sight of Right Eye necessary for Militia Recruits.—It is laid down in the 'Regulations for the Militia, War Office, 1883,' as regards the medical examination of recruits (Part 1, sect. 2, page 32), that one of the principal points to be attended to is, ' that his (the recruit's) vision is good, or at least sufficiently good to enable him with his right eye to discern objects clearly at not less than 300 yards.' No particular size of object is specified in these Regulations, nor is any visual limit mentioned as regards the left eye.

Indian Government Services.—The rules affecting candidates for the various branches of the Indian Government Service have been already mentioned under their separate headings in Ch. IX. They are explained in detail in a pamphlet entitled ' Regulations as to Defects of Vision which Disqualify Candidates for Admission into the Civil or Military Government Services. By Sir J. Fayrer, K.C.S.I. LL.D., F.R.S., &c., President of the Medical Board, India Office. 2nd Edition. London: J. and A. Churchill, 1887.'

Quality of Vision necessary for Officers and Men of the Royal Navy and Royal Marines.—The following directions to medical officers in regard to the examination of the quality of vision possessed by men or boys seeking admission into the Naval Service or into the Royal Marines are laid down in the ' Queen's Regulations and Admiralty Instructions (1879).' The medical officer is ordered to observe whether the person's eyesight is defective, and if it should be so to such an extent as might, in the opinion of the examining officer, disqualify him for the efficient discharge of the duties that would devolve on him, he is to report him unfit for the service. He is further instructed (par. 1074) that ' the eyes should be clear, intelligent, expressive of health, and the eyesight good.' In the succeeding paragraph of the ' Instructions ' (par. 1075), it is laid down that ' persons of whatever class or age, who are found to be labouring under certain physical defects, are to be considered unfit for Her Majesty's Service,' and among these defects is mentioned ' blindness or defective vision in one or both eyes.' The following directions also occur in the same code of ' Regulations ': ' Whenever test-types are supplied, the power of vision of each eye separately, as well as together, is to be ascertained; but before finally rejecting a person who has failed to read the types, he is to be tested with objects familiar to him, and at distances corresponding to the sizes of the objects, as the inability occasionally arises from other " causes than defective " sight. When the sight is found defective, the particulars are to be recorded in the register of physical examination when required to be kept, how far short of the normal distances given in the test-types could the letters or figures be seen, and, where one is more defective than the other, the limit of good vision of each eye is to be noted.'

It is all the more important that naval officers and seamen should possess normal acuteness of vision, inasmuch as the use of correcting glasses is less admissible under the conditions of sea life than it is among officers and men engaged in ordinary military duties on land.

Rules for determining Visual Competency in the German Navy.—The fitness of candidates in the Navy of Germany is determined by the degree of visual acuteness, and the capability of its correction by suitable spectacles in case it is below the normal standard. The Imperial Admiralty Orders (June 26, 1872) are to the following effect :—

1. The acuteness of vision is to be tested by Snellen's test-types. When the types can be read at the denominated distances it is to be considered normal, or = 1. The certificate must state clearly the result of this examination.

2. If the acuteness of vision be not normal, the surgeon must determine by ophthalmoscopic examination whether there is any organic disease of the inner parts of the eye. If there be, the candidate is to be considered unfit.

3. In the absence of organic disease of the eye, the following limits are to be adhered to :—

    (a) Candidates who recognise Snellen's types at $\frac{3}{4}$ of the denominated distance, i.e. whose visual acuteness is $= \frac{3}{4}$, are to be considered fit for the naval service.

    (b) Candidates who recognise the types at distances between $\frac{3}{4}$ and $\frac{1}{2}$ of the normal distance, can be admitted, provided it is proved, by the application of spectacles, that their diminished visual power is perfectly corrected by their help.

    (c) Candidates whose visual power is only $\frac{1}{2}$ or less, are to be considered unfit for the naval service.

Degree of M., Deficiency of Visual Acuteness, etc., which exclude from Military Service in Foreign Armies.—The degrees of ametropia and amblyopia which exclude men from military service, or which exclude from service in particular parts of foreign armies, vary in different countries. They have been considerably modified in most of them of late years, and the rules shown in previous editions of this Manual are now altered in many particulars. In some countries the rules regarding the degrees of the disqualifying defects, and also regarding the methods by which they are to be determined, are now laid down with much minuteness and precision. A study of the various regulations regarding visual defects and the modes of testing them will be found to be very instructive. I have obtained copies of the most recent orders on these subjects in the principal armies of Europe, and in collecting them Dr. Gori, Lector on Military Surgery and Hygiene in the University of Amsterdam, has afforded me valuable assistance.

Holland.—In Holland, Dr. Gori informs me, the Royal Act of November, 1883, decrees that M. = 2·5 D and upwards of the right eye in case the left eye is normal, and 7 D and upwards of the left eye in case the right eye is normal, renders a conscript unfit for service. The degree of M. is to be determined after mydriasis by atropine (homatropini hydrobrom. 2 per cent.), or by a refraction ophthalmoscope.

H. (total hypermetropia) = 6·0 D of right eye, in case the left eye is normal, and H. = 9·0 D, or higher of the left eye, in case the right eye is normal, also unfit for military service.

Ast. to such an amount that V. (visual acuteness) is reduced to less than $\frac{1}{4}$th in the right eye, the left being normal, or to less than $\frac{1}{20}$th in the left eye, the right eye being normal, entails unfitness for service in the army.

The rules are different for the volunteers of the military service.

By the Act already quoted, volunteers using portable firearms must have V. (visual acuteness), without glasses, not less than $\frac{3}{4}$ths for the right eye, or $\frac{1}{2}$ for the left eye.

In case volunteers cannot fulfil the above requirement, owing to M. or H., they are admissible for service, if under 20 years of age, with M. $= 1$ D, or if over that age with M. $= 1\cdot5$ D; and with Hm. $= 1\cdot0$ D if under 20 years of age, and with Hm. $= 2\cdot0$ D if beyond that age; provided that V. after correction by glasses is rendered distinctly $= 1$.

Medical cadets and candidates for army commissions who do not require acute distant vision are admissible if V. of one eye is not less than $\frac{3}{4}$ths, and of the other not less than $\frac{1}{2}$; or with V. of one eye equal to 1, and V. not below $\frac{1}{3}$rd of the other. In case of M. being the cause of the deficiency of V., medical cadets and candidates are admissible with M. $= 3\cdot0$ D, or, in case of H. being the cause, with H. not exceeding $2\cdot0$ D, provided V. is rendered $= 1$ by correction with glasses.

France.—In accordance with the instructions of the French Army Sanitary Council (Conseil de Santé des Armées), of February 27, 1877, on the diseases or faults of conformation which unfit for military service, whatever the nature of the lesion may be, if it reduces V. to $\frac{1}{4}$th in the two eyes, or in the right eye, or to $\frac{1}{12}$th in the left eye, or causes a diminution of about a half of the temporal angle of the field of vision, it renders the subject unfit for military service in the French army, unless, being the result of simply refractive defect, it can be corrected by glasses.

M., however, higher than $6\cdot0$ D, or complicated with muscular or accommodatory insufficiency, or with lesion of the ocular fundus, renders the subject of it unfit for service in the army. In the auxiliary service, M. between $6\cdot0$ D and $9\cdot0$ D is a cause of unfitness. The degree of M. must be determined by the optometer or by the ophthalmoscope.

H. renders unfit for service whenever it causes V. to be below $\frac{1}{4}$th in the right, or $\frac{1}{12}$th in the left eye. The determination of H. suffices; it is not necessary to specify the degree. The upright image of the fundus must be clearly visible, without dilatation of the pupil, by the aid of an ophthalmoscopic mirror at a distance of from 4 to 6 inches from the eye. In the auxiliary service H. which lowers V. below $\frac{1}{4}$th is a cause of unfitness, even though capable of correction by glasses.

Ast., like H., renders unfit for service whenever it causes V. to be below $\frac{1}{4}$th in the right, and $\frac{1}{12}$th in the left eye.

By decree of May 12, 1877, the use of spectacles by soldiers in the ranks was permitted, and in 1879 a certain number of pairs of spectacles were officially sent to different corps of the army.

Germany.—In the German army a reduction of V. to $\frac{1}{4}$th, or below, causes permanent unfitness for military service in recruits;

reduction of V. in both eyes to $\frac{1}{2}$, or less, but not as low as $\frac{1}{4}$th, allows recruits to be conditionally fit.   M., when the distant point of the better eye is 0·15 m. (6 in.), or less, even though V. is normal at and within this distance, renders recruits permanently unfit; M., with a distant point of more than 0·15 m., when V. after correction by glasses is more than half normal V., allows conditional fitness.   Blindness of either eye is a cause of unfitness. Examination of V. in the recruiting service is, as a rule, to be by sight-tests, and Snellen's test-types are to be used till further orders.   The results are to be expressed as regards V. in non-reduced fractions ($\frac{10}{40}$ or $\frac{20}{40}$, not $\frac{1}{4}$ or 1).

As regards drilled soldiers, and men entitled to pension, reduction of V. in both eyes, if less than $\frac{1}{2}$, but more than $\frac{1}{4}$, of normal V., causes unfitness for field service ; reduction of V. in the better of the two eyes to $\frac{1}{4}$ or less causes unfitness both for field and garrison service.   Blindness of one eye causes also unfitness for field and garrison service.

No special rules are laid down as regards H. or Ast.

A War Ministry Order of July 15, 1881, makes known to medical officers that garrison-hospitals can be supplied with bi-concave or bi-convex spectacles for troops of the infantry, and spectacles with thread-covered steel frames for mounted troops, in a case at $2\frac{1}{2}$ marks (about 2s. 6d.), or, with postage, 2·9 marks (nearly 3s.).

Italy.—The Royal Decree of September 26, 1881, by which the previous list of infirmities exempting from military service was modified, rules that reduction of V. to $\frac{1}{3}$rd of the normal in the right eye, or to $\frac{1}{12}$th in the left eye, although V. in the right eye is up to the normal, if caused by organic changes or incurable disorders of the globe of the eye, incapacitates for military service in the army of Italy.   The existence and degrees of the defects must be established in a military hospital, must be recognised to be irremediable, and their nature must be specified by a medical officer skilled in their diagnosis.   By normal vision is understood the V. which permits objects to be distinguished under a visual angle not larger than 5′.

M. = 6·0 D or higher in the right eye, when accommodation is paralysed, incapacitates for military service.

Hm., such that with Acc. intact, and with the naked eye, there is not the power, under binocular vision, and at a distance of 0·30 m. (12 inches), to read printed characters of one millimetre in height, or to distinguish signs and objects of like dimensions; and that, with Acc. paralysed (total H.), amounts to 6·0 D in the right eye, exempts from military service.

Ast., such that V. in the right eye is reduced below $\frac{1}{3}$rd of normal V., also renders unfit for military service.

The existence and degrees of the refractive defects above

mentioned must be certified in a military hospital after employing all the scientific means available, including ophthalmoscopic observation and, when needed, atropinisation.

As regards refractive defect, whatever may be its degree, if the left eye only is involved, it does not cause exemption from military service.

Austria.—The Austrian Recruiting Regulations of 1883 ('Instruktion zur aerztlichen Untersuchung der Wehrpflichtigen,' Wien, 1883) give more precise directions as regards visual limits of fitness for military service than previous regulations on the subject. The necessary visual power (V.), which was before at the discretion of individual examiners, is fixed by them. It is mentioned in them that Snellen's types and objects, metrically numbered, have been introduced as sight-tests.

Reduction of V., but leaving more than one-half of normal V., does not incapacitate men for military service who are fit in other respects; reduction of V., but leaving more than $\frac{1}{2}$ normal V. in the right eye, accompanied by reduction of V. to less than $\frac{1}{2}$, or even to $\frac{1}{4}$th of normal V. in the left eye, incapacitates for general military service, but does not incapacitate for the Purveying and Clothing Departments and certain other portions of the army; reduction of V. to less than $\frac{1}{2}$ in the better of the two eyes incapacitates for military service, but only after subsequent confirmation.

M. with a distant point of distinct V. at 12 inches from the eye, or beyond, does not cause unfitness for military service. M. with a distant point limited to 8 inches, if V. is good, though rendering unfit for general military service, does not incapacitate for service as medical officers, dispensers, in the clothing and administrative departments, or as one-year volunteers. M. with a far point of 12 inches or less, if the man is able to read printed letters, or recognise figures of $\frac{1}{3}$rd of a Vienna line in height and breadth at any distance from the eye when wearing concave 4-inch spectacles, determines unfitness for military service. If the man is able to read the test-objects under these conditions, he gives a positive proof that he is myopic beyond the limit of fitness, i.e. beyond $\frac{1}{12}$th. If, although myopic, the man cannot read the print with the $-4''$ spectacles, he is to be sent to a military hospital for further scientific examination.

H. of such a degree that the man is able to read printed letters, or recognise other characters of 1 Vienna line in height, and of corresponding breadth, at a distance of more than 12 inches from the eye when wearing $+6''$ spectacles, unfits for service. Experience has proved that hypermetropic persons can comply with this test only when their H. is above $\frac{1}{4}$th. If the hypermetrope cannot satisfy this test, further special examination is to be made at a military hospital. If H. is not so high in degree as $\frac{1}{4}$th, it does not cause unfitness for military service.

Strabismus affecting either eye, if its central visual power is less than $\frac{1}{4}$ of normal V., determines unfitness; strabismus affecting the left eye, if its visual power is more than $\frac{1}{4}$ of normal V., does not cause unfitness for military service.

Belgium.—The latest Belgian Regulations determine that reduction of V. in the right eye to $\frac{1}{3}$rd of normal V. causes exemption from military service. Reduction of V. even below this limit in the left eye, by itself, does not cause exemption. Blindness of either eye causes unfitness for military service in the Belgian army.

M. of right eye, after paralysis of Acc., = 6·0 D or above, causes unfitness for military service. M. of even higher degree, in the left eye, does not by itself warrant exemption from service.

H. of right eye, after paralysis of Acc., = 6·0 D or above, causes unfitness for military service. H. of even higher degree in the left eye does not by itself warrant exemption from service.

Strabismus, with considerable contraction of the visual field, exempts from military service.

Switzerland.—The Swiss Army Regulations lay down the rules that the minimum of V. for the artillery is 1 ; for the infantry is $\frac{2}{3}$ths ; and for all other troops is $\frac{1}{2}$ of normal V. All men in whom deficiency of V. may be corrected by ordinary + or − spherical glasses, and brought to a standard of $\frac{1}{2}$ up to normal V., are ruled to be fit for military service. They are permitted to wear spectacles. V. less than $\frac{1}{2}$ incapacitates for service, excepting as regards medical officers, in whom V. = $\frac{1}{3}$ is tolerated. Whenever V. in one eye is normal, V. in the other may be as low as $\frac{1}{3}$th ; but, unless the eye with normal V. is the right eye, such men cannot be employed as riflemen.

Either M. or H. of higher degree than 4·0 D excludes from service in infantry or cavalry, even although it admits of correction by glasses.

Ast. is judged under the general rules for V.

Denmark.—The rules in the Danish army are the following :—

V. below $\frac{1}{6}$th ($\frac{20}{100}$ Snellen) incapacitates for military service.

M. up to 2·25 |D for combatants, and from 2·50 D to 5·00 D for other troops, does not incapacitate for military service; above 5·00 D unfits for service.

H. up to 4·50 D for combatants, above this degree for other troops, incapacitates for military service.

Ast. is judged by the standard of visual acuteness.

Spain and Portugal.—In the Portuguese army, M. of such a degree that the man can read or distinguish small objects at a distance of 0·25 m. (10 in.) with glasses of − 8·0 D, or higher, and can see distant objects through glasses of − 7·0 D, unfits for military service. In the Spanish army M., characterised by the possibility of reading small printed characters at a distance of 0·35 m. (14 in.) with glasses of − 6·0 D, and being unable to distinguish them with − 2·0 D, exempts from military service. No precise rules appear to be laid down as regards H. or Ast. in the Regulations of either the Spanish or Portuguese armies.

INDEX.

PRINTED BY
SPOTTISWOODE AND CO., NEW-STREET SQUARE
LONDON

February 1888.

Catalogue of Books

PUBLISHED BY

MESSRS. LONGMANS, GREEN, & CO.

39 PATERNOSTER ROW, LONDON, E.C.

Abbey.—*THE ENGLISH CHURCH AND ITS BISHOPS*, 1700–1800. By CHARLES J. ABBEY, Rector of Checkendon. 2 vols. 8vo. 24*s.*

Abbey and Overton.—*THE ENGLISH CHURCH IN THE EIGHTEENTH CENTURY.* By CHARLES J. ABBEY, Rector of Checkendon, and JOHN H. OVERTON, Rector of Epworth and Canon of Lincoln. Crown 8vo. 7*s. 6d.*

Abbott.—*THE ELEMENTS OF LOGIC.* By T. K. ABBOTT, B.D. 12mo. 3*s.*

Acton. — *MODERN COOKERY FOR PRIVATE FAMILIES.* By ELIZA ACTON. With 150 Woodcuts. Fcp. 8vo. 4*s. 6d.*

Adams.—*PUBLIC DEBTS:* an Essay on the Science of Finance. By HENRY C. ADAMS, Ph.D. 8vo. 12*s. 6d.*

A. K. H. B.—*THE ESSAYS AND CONTRIBUTIONS OF A. K. H. B.*—Uniform Cabinet Editions in crown 8vo.
Autumn Holidays of a Country Parson, 3*s. 6d.*
Changed Aspects of Unchanged Truths, 3*s. 6d.*
Commonplace Philosopher, 3*s. 6d.*
Counsel and Comfort from a City Pulpit, 3*s. 6d.*
Critical Essays of a Country Parson, 3*s. 6d.*
Graver Thoughts of a Country Parson. Three Series, 3*s. 6d.* each.
Landscapes, Churches, and Moralities, 3*s. 6d.*
Leisure Hours in Town, 3*s. 6d.*
Lessons of Middle Age, 3*s. 6d.*
Our Little Life. Two Series, 3*s. 6d.* each.
Our Homely Comedy and Tragedy, 3*s. 6d.*
Present Day Thoughts, 3*s. 6d.*
Recreations of a Country Parson. Three Series, 3*s. 6d.* each.
Seaside Musings, 3*s. 6d.*
Sunday Afternoons in the Parish Church of a Scottish University City, 3*s. 6d.*

Amos.—*WORKS BY SHELDON AMOS.*
A PRIMER OF THE ENGLISH CONSTITUTION AND GOVERNMENT. Crown 8vo. 6*s.*
A SYSTEMATIC VIEW OF THE SCIENCE OF JURISPRUDENCE. 8vo. 18*s.*

Aristotle.—*THE WORKS OF.*
THE POLITICS, G. Bekker's Greek Text of Books I. III. IV. (VII.) with an English Translation by W. E. BOLLAND, M.A.; and short Introductory Essays by A. LANG, M.A. Crown 8vo. 7*s. 6d.*

THE POLITICS: Introductory Essays. By ANDREW LANG. (From Bolland and Lang's 'Politics.') Crown 8vo. 2*s. 6d.*

THE ETHICS; Greek Text, illustrated with Essays and Notes. By Sir ALEXANDER GRANT, Bart. M.A. LL.D. 2 vols. 8vo. 32*s.*

THE NICOMACHEAN ETHICS, Newly Translated into English. By ROBERT WILLIAMS, Barrister-at-Law. Crown 8vo. 7*s. 6d.*

Armstrong.—*WORKS BY GEORGE FRANCIS ARMSTRONG, M.A.*
POEMS: Lyrical and Dramatic. Fcp. 8vo. 6*s.*
KING SAUL. (The Tragedy of Israel, Part I.) Fcp. 8vo. 5*s.*
KING DAVID. (The Tragedy of Israel, Part II.) Fcp. 8vo. 6*s.*
KING SOLOMON. (The Tragedy of Israel, Part III.) Fcp. 8vo. 6*s.*
UGONE: A Tragedy. Fcp. 8vo. 6*s.*
A GARLAND FROM GREECE; Poems. Fcp. 8vo. 9*s.*
STORIES OF WICKLOW; Poems. Fcp. 8vo. 9*s.*
VICTORIA REGINA ET IMPERATRIX: a Jubilee Song from Ireland, 1887. 4to. 5*s.* cloth gilt.
THE LIFE AND LETTERS OF EDMUND J. ARMSTRONG. Fcp. 8vo. 7*s. 6d.*

Armstrong.—*WORKS BY EDMUND J. ARMSTRONG.*
POETICAL WORKS. Fcp. 8vo. 5*s.*
ESSAYS AND SKETCHES. Fcp. 8vo. 5*s.*

Arnold. — *WORKS BY THOMAS ARNOLD, D.D. Late Head master of Rugby School.*

INTRODUCTORY LECTURES ON MODERN HISTORY, delivered in 1841 and 1842. 8vo. 7s. 6d.

SERMONS PREACHED MOSTLY IN THE CHAPEL OF RUGBY SCHOOL. 6 vols. crown 8vo. 30s. or separately, 5s. each.

MISCELLANEOUS WORKS. 8vo. 7s. 6d.

Arnold.—*A MANUAL OF ENGLISH LITERATURE,* Historical and Critical. By THOMAS ARNOLD, M.A. Crown 8vo. 7s. 6d.

Arnott.—*THE ELEMENTS OF PHYSICS OR NATURAL PHILOSOPHY.* By NEIL ARNOTT, M.D. Edited by A. BAIN, LL.D. and A. S. TAYLOR, M.D. F.R.S. Woodcuts. Crown 8vo. 12s. 6d.

Ashby. — *NOTES ON PHYSIOLOGY FOR THE USE OF STUDENTS PREPARING FOR EXAMINATION.* With 120 Woodcuts. By HENRY ASHBY, M.D. Lond. Fcp. 8vo. 5s.

Atelier (The) du Lys; or, an Art Student in the Reign of Terror. By the Author of 'Mademoiselle Mori.' Crown 8vo. 2s. 6d.

Bacon.—*THE WORKS AND LIFE OF.*

COMPLETE WORKS. Edited by R. L. ELLIS, M.A. J. SPEDDING, M.A. and D. D. HEATH. 7 vols. 8vo. £3. 13s. 6d.

LETTERS AND LIFE, INCLUDING ALL HIS OCCASIONAL WORKS. Edited by J. SPEDDING. 7 vols. 8vo. £4. 4s.

THE ESSAYS; with Annotations. By RICHARD WHATELY, D.D., 8vo. 10s. 6d.

THE ESSAYS; with Introduction, Notes, and Index. By E. A. ABBOTT, D.D. 2 vols. fcp. 8vo. price 6s. Text and Index only, without Introduction and Notes, in 1 vol. fcp. 8vo. 2s. 6d.

Bagehot.—*WORKS BY WALTER BAGEHOT, M.A.*

BIOGRAPHICAL STUDIES. 8vo. 12s.

ECONOMIC STUDIES. 8vo. 10s. 6d.

LITERARY STUDIES. 2 vols. 8vo. 28s.

THE POSTULATES OF ENGLISH POLITICAL ECONOMY. Crown 8vo. 2s. 6d.

The BADMINTON LIBRARY, edited by the DUKE OF BEAUFORT, K.G. assisted by ALFRED E. T. WATSON.

Hunting. By the DUKE OF BEAUFORT, K.G. and MOWBRAY MORRIS. With Contributions by the Earl of Suffolk and Berkshire, Rev. E. W. L. Davies, Digby Collins, and Alfred E. T. Watson. With Coloured Frontispiece and 53 Illustrations by J. Sturgess, J. Charlton, and Agnes M. Biddulph. Crown 8vo. 10s. 6d.

Fishing. By H. CHOLMONDELEY-PENNELL. With Contributions by the Marquis of Exeter, Henry R. Francis, M.A., Major John P. Traherne, G. Christopher Davies, R. B. Marston, &c.

Vol. I. Salmon, Trout, and Grayling. With 150 Illustrations. Cr. 8vo. 10s. 6d.

Vol. II. Pike and other Coarse Fish. With 58 Illustrations. Cr. 8vo. 10s. 6d.

Racing and Steeplechasing. By the EARL OF SUFFOLK AND BERKSHIRE, W. G. CRAVEN, The Hon. F. LAWLEY, A. COVENTRY, and A. E. T. WATSON. With Coloured Frontispiece and 56 Illustrations by J. Sturgess. Cr. 8vo. 10s. 6d.

Shooting. By Lord WALSINGHAM and Sir RALPH PAYNE-GALLWEY, Bart. with Contributions by Lord Lovat, Lord Charles Lennox Kerr, The Hon. G. Lascelles, and Archibald Stuart Wortley. With 21 full-page Illustrations and 149 Woodcuts by A. J. Stuart-Wortley, C. Whymper, J. G. Millais, &c.

Vol. I. Field and Covert. Cr. 8vo. 10s. 6d.

Vol. II. Moor and Marsh. Cr. 8vo. 10s. 6d.

Cycling. By VISCOUNT BURY, K.C.M.G. and G. LACY HILLIER. With 19 Plates and 61 Woodcuts by Viscount Bury and Joseph Pennell. Cr. 8vo. 10s. 6d.

Athletics and Football. By MONTAGUE SHEARMAN. With Introduction by Sir Richard Webster, Q.C. M.P. With 6 full-page Illustrations and 45 Woodcuts from Drawings by Stanley Berkeley, and from Instantaneous Photographs by G. Mitchell. Cr. 8vo. 10s. 6d.

*** Other volumes in preparation.

Bagwell. — *IRELAND UNDER THE TUDORS,* with a Succinct Account of the Earlier History. By RICHARD BAGWELL, M.A. Vols. I. and II. From the first invasion of the Northmen to the year 1578. 2 vols. 8vo. 32s.

Bain. — WORKS BY ALEXANDER BAIN, LL.D.

MENTAL AND MORAL SCIENCE; a Compendium of Psychology and Ethics. Crown 8vo. 10s. 6d.

THE SENSES AND THE INTELLECT. 8vo. 15s.

THE EMOTIONS AND THE WILL. 8vo. 15s.

PRACTICAL ESSAYS. Cr. 8vo. 4s. 6d.

LOGIC, DEDUCTIVE AND INDUCTIVE. PART I. Deduction, 4s. PART II. Induction, 6s. 6d.

JAMES MILL; a Biography. Cr. 8vo. 2s.

JOHN STUART MILL; a Criticism, with Personal Recollections. Cr. 8vo. 1s.

Baker. — WORKS BY SIR SAMUEL W. BAKER, M.A.

EIGHT YEARS IN CEYLON. Crown 8vo. Woodcuts. 5s.

THE RIFLE AND THE HOUND IN CEYLON. Crown 8vo. Woodcuts. 5s.

Bale. — A HANDBOOK FOR STEAM USERS; being Notes on Steam Engine and Boiler Management and Steam Boiler Explosions. By M. POWIS BALE, M.I.M.E. A.M.I.C.E. Fcp. 8vo. 2s. 6d.

Ball. — THE REFORMED CHURCH OF IRELAND (1537-1886). By the Right Hon. J. T. BALL, LL.D. D.C.L. 8vo. 7s. 6d.

Barker. — A SHORT MANUAL OF SURGICAL OPERATIONS, having Special Reference to many of the Newer Procedures. By A. E. J. BARKER, F.R.C.S. Surgeon to University College Hospital. With 61 Woodcuts. Crown 8vo. 12s. 6d.

Barrett. — ENGLISH GLEES AND PART-SONGS. An Inquiry into their Historical Development. By WILLIAM. ALEXANDER BARRETT. 8vo. 7s. 6d.

Beaconsfield. — WORKS BY THE EARL OF BEACONSFIELD, K.G.

NOVELS AND TALES. The Hughenden Edition. With 2 Portraits and 11 Vignettes. 11 vols. Crown 8vo. 42s.

Endymion.

Lothair.	Henrietta Temple.
Coningsby.	Contarini Fleming, &c.
Sybil.	Alroy, Ixion, &c.
Tancred.	The Young Duke, &c.
Venetia.	Vivian Grey.

NOVELS AND TALES. Cheap Edition, complete in 11 vols. Crown 8vo. 1s. each, boards; 1s. 6d. each, cloth.

THE WIT AND WISDOM OF THE EARL OF BEACONSFIELD. Crown 8vo. 1s. boards, 1s. 6d. cloth.

Becker. — WORKS BY PROFESSOR BECKER, translated from the German by the Rev. F. METCALF.

GALLUS; or, Roman Scenes in the Time of Augustus. Post 8vo. 7s. 6d.

CHARICLES; or, Illustrations of the Private Life of the Ancient Greeks. Post 8vo. 7s. 6d.

Bentley. — A TEXT-BOOK OF ORGANIC MATERIA MEDICA. Comprising a Description of the VEGETABLE and ANIMAL DRUGS of the BRITISH PHARMACOPŒIA, with some others in common use. Arranged Systematically and especially Designed for Students. By ROBT. BENTLEY, M.R.C.S. Eng. F.L.S. With 62 Illustrations. Crown 8vo. 7s. 6d.

Boultbee. — A COMMENTARY ON THE 39 ARTICLES of the Church of England. By the Rev. T. P. BOULTBEE, LL.D. Crown 8vo. 6s.

Bourne. — WORKS BY JOHN BOURNE, C.E.

CATECHISM OF THE STEAM ENGINE in its various Applications in the Arts, to which is now added a chapter on Air and Gas Engines, and another devoted to Useful Rules, Tables, and Memoranda. Illustrated by 212 Woodcuts. Crown 8vo. 7s. 6d.

HANDBOOK OF THE STEAM ENGINE; a Key to the Author's Catechism of the Steam Engine. With 67 Woodcuts. Fcp. 8vo. 9s.

RECENT IMPROVEMENTS IN THE STEAM ENGINE. With 124 Woodcuts. Fcp. 8vo. 6s.

Bowen. — HARROW SONGS AND OTHER VERSES. By EDWARD E. BOWEN. Fcp. 8vo. 2s. 6d.; or printed on hand-made paper, 5s.

Brassey. — WORKS BY LADY BRASSEY.

A VOYAGE IN THE 'SUNBEAM,' OUR HOME ON THE OCEAN FOR ELEVEN MONTHS.

Library Edition. With 8 Maps and Charts, and 118 Illustrations, 8vo. 21s.

Cabinet Edition. With Map and 66 Illustrations, crown 8vo. 7s. 6d.

School Edition. With 37 Illustrations, fcp. 2s. cloth, or 3s. white parchment with gilt edges.

Popular Edition. With 60 Illustrations, 4to. 6d. sewed, 1s. cloth.

[Continued on next page.

Brassey. — WORKS BY LADY BRASSEY—*continued.*

SUNSHINE AND STORM IN THE EAST.
Library Edition. With 2 Maps and 114 Illustrations, 8vo. 21s.
Cabinet Edition. With 2 Maps and 114 Illustrations, crown 8vo. 7s. 6d.
Popular Edition. With 103 Illustrations, 4to. 6d. sewed, 1s. cloth.

IN THE TRADES, THE TROPICS, AND THE 'ROARING FORTIES.'
Cabinet Edition. With Map and 220 Illustrations, crown 8vo. 7s. 6d.
Popular Edition. With 183 Illustrations, 4to. 6d. sewed, 1s. cloth.

THREE VOYAGES IN THE 'SUNBEAM.'
Popular Edition. With 346 Illustrations, 4to. 2s. 6d.

Browne.—AN EXPOSITION OF THE 39 ARTICLES, Historical and Doctrinal. By E. H. BROWNE, D.D., Bishop of Winchester. 8vo. 16s.

Bryant.—EDUCATIONAL ENDS; or, the Ideal of Personal Development. By SOPHIE BRYANT, D.Sc.Lond. Crown 8vo. 6s.

Buckle. — WORKS BY HENRY THOMAS BUCKLE.

HISTORY OF CIVILISATION IN ENGLAND AND FRANCE, SPAIN AND SCOTLAND. 3 vols. crown 8vo. 24s.

MISCELLANEOUS AND POSTHUMOUS WORKS. ·A New and Abridged Edition. Edited by GRANT ALLEN. 2 vols. crown 8vo. 21s.

Buckton.—WORKS BY MRS. C. M. BUCKTON.

FOOD AND HOME COOKERY. With 11 Woodcuts. Crown 8vo. 2s. 6d.

HEALTH IN THE HOUSE. With 41 Woodcuts and Diagrams. Crown 8vo. 2s.

OUR DWELLINGS. With 39 Illustrations. Crown 8vo. 3s. 6d.

Bull.—WORKS BY THOMAS BULL, M.D.

HINTS TO MOTHERS ON THE MANAGEMENT OF THEIR HEALTH during the Period of Pregnancy and in the Lying-in Room. Fcp. 8vo. 1s. 6d.

THE MATERNAL MANAGEMENT OF CHILDREN IN HEALTH AND DISEASE. Fcp. 8vo. 1s. 6d.

Bullinger.—A CRITICAL LEXICON AND CONCORDANCE TO THE ENGLISH AND GREEK NEW TESTAMENT. Together with an Index of Greek Words and several Appendices. By the Rev. E. W. BULLINGER, D.D. Royal 8vo. 15s.

Burrows.—THE FAMILY OF BROCAS OF BEAUREPAIRE AND ROCHE COURT, Hereditary Masters of the Royal Buckhounds. With some account of the English Rule in Aquitaine. By MONTAGU BURROWS, M.A. F.S.A. With 26 Illustrations of Monuments, Brasses, Seals, &c. Royal 8vo. 42s.

Cabinet Lawyer, The; a Popular Digest of the Laws of England, Civil, Criminal, and Constitutional. Fcp. 8vo. 9s.

Canning.—SOME OFFICIAL CORRESPONDENCE OF GEORGE CANNING. Edited, with Notes, by EDWARD J. STAPLETON. 2 vols. 8vo. 28s.

Carlyle. — THOMAS AND JANE WELSH CARLYLE.

THOMAS CARLYLE, a History of the first Forty Years of his Life, 1795–1835. By J. A. FROUDE, M.A. With 2 Portraits and 4 Illustrations, 2 vols. 8vo. 32s.

THOMAS CARLYLE, a History of his Life in London : from 1834 to his death in 1881. By J. A. FROUDE, M.A. 2 vols. 8vo. 32s.

LETTERS AND MEMORIALS OF JANE WELSH CARLYLE. Prepared for publication by THOMAS CARLYLE, and edited by J. A. FROUDE, M.A. 3 vols. 8vo. 36s.

Cates. — A DICTIONARY OF GENERAL BIOGRAPHY. Fourth Edition, with Supplement brought down to the end of 1884. By W. L. R. CATES. 8vo. 28s. cloth ; 35s. half-bound russia.

Clerk.—THE GAS ENGINE. By DUGALD CLERK. With 101 Illustrations and Diagrams. Crown 8vo. 7s. 6d.

Clodd.—THE STORY OF CREATION: a Plain Account of Evolution. By EDWARD CLODD, Author of 'The Childhood of the World' &c. With 77 Illustrations. Crown 8vo. 6s.

Coats.—A MANUAL OF PATHOLOGY. By JOSEPH COATS, M.D. Pathologist to the Western Infirmary and the Sick Children's Hospital, Glasgow. With 339 Illustrations engraved on Wood. 8vo. 31s. 6d.

Colenso.—*THE PENTATEUCH AND BOOK OF JOSHUA CRITICALLY EXAMINED.* By J. W. COLENSO, D.D. late Bishop of Natal. Crown 8vo. 6s.

Comyn.—*ATHERSTONE PRIORY:* a Tale. By L. N. COMYN. Crown 8vo. 2s. 6d.

Conder. — *A HANDBOOK TO THE BIBLE,* or Guide to the Study of the Holy Scriptures derived from Ancient Monuments and Modern Exploration. By F. R. CONDER, and Lieut. C. R. CONDER, R.E. Post 8vo. 7s. 6d.

Conington. — *WORKS BY JOHN CONINGTON, M.A.*

THE ÆNEID OF VIRGIL. Translated into English Verse. Crown 8vo. 9s.

THE POEMS OF VIRGIL. Translated into English Prose. Crown 8vo. 9s.

Conybeare & Howson. — *THE LIFE AND EPISTLES OF ST. PAUL.* By the Rev. W. J. CONYBEARE, M.A. and the Very Rev. J. S. HOWSON, D.D.

Library Edition, with Maps, Plates, and Woodcuts. 2 vols. square crown 8vo. 21s.

Student's Edition, revised and condensed, with 46 Illustrations and Maps. 1 vol. crown 8vo. 7s. 6d.

Cooke. — *TABLETS OF ANATOMY.* By THOMAS COOKE, F.R.C.S. Eng. B.A. B.Sc. M.D. Paris. Fourth Edition, being a selection of the Tablets believed to be most useful to Students generally. Post 4to. 7s. 6d.

Cox. — *THE FIRST CENTURY OF CHRISTIANITY.* By HOMERSHAM COX, M.A. 8vo. 12s.

Cox.—*A GENERAL HISTORY OF GREECE:* from the Earliest Period to the Death of Alexander the Great; with a Sketch of the History to the Present Time. By the Rev. Sir G. W. Cox, Bart., M.A. With 11 Maps and Plans. Crown 8vo. 7s. 6d.

*** For other Works by Sir G. Cox, see 'Epochs of History,' p. 24.

Creighton. — *HISTORY OF THE PAPACY DURING THE REFORMATION.* By the Rev. M. CREIGHTON, M.A. 8vo. Vols. I. and II. 1378–1464, 32s.; Vols. III. and IV. 1464–1518, 24s.

Crookes. — *SELECT METHODS IN CHEMICAL ANALYSIS* (chiefly Inorganic). By WILLIAM CROOKES, F.R.S. V.P.C.S. With 37 Illustrations. 8vo. 24s.

Crozier.—*CIVILIZATION AND PROGRESS.* By JOHN BEATTIE CROZIER. New and Cheaper Edition. 8vo. 5s.

Crump.—*A SHORT ENQUIRY INTO THE FORMATION OF POLITICAL OPINION,* from the Reign of the Great Families to the Advent of Democracy. By ARTHUR CRUMP. 8vo. 7s. 6d.

Culley.—*HANDBOOK OF PRACTICAL TELEGRAPHY.* By R. S. CULLEY, M. Inst. C.E. Plates and Woodcuts. 8vo. 16s.

Dante.—*THE DIVINE COMEDY OF DANTE ALIGHIERI.* Translated verse for verse from the Original into Terza Rima. By JAMES INNES MINCHIN. Crown 8vo. 15s.

Davidson.—*AN INTRODUCTION TO THE STUDY OF THE NEW TESTAMENT,* Critical, Exegetical, and Theological. By the Rev. S. DAVIDSON, D.D. LL.D. Revised Edition. 2 vols. 8vo. 30s.

Davidson.—*WORKS BY WILLIAM L. DAVIDSON, M.A.*

THE LOGIC OF DEFINITION EXPLAINED AND APPLIED. Crown 8vo. 6s.

LEADING AND IMPORTANT ENGLISH WORDS EXPLAINED AND EXEMPLIFIED. Fcp. 8vo. 3s. 6d.

Decaisne & Le Maout. — *A GENERAL SYSTEM OF BOTANY.* Translated from the French of E. LE MAOUT, M.D., and J. DECAISNE, by Mrs. HOOKER; with Additions by Sir J. D. HOOKER, C.B. F.R.S. Imp. 8vo. with 5,500 Woodcuts, 31s. 6d.

De Salis. — *WORKS BY MRS. DE SALIS.*

SAVOURIES À LA MODE. Fcp. 8vo. 1s. boards.

ENTRÉES À LA MODE. Fcp. 8vo. 1s. 6d. boards.

SOUPS AND DRESSED FISH À LA MODE. Fcp. 8vo. 1s. 6d. boards.

OYSTERS À LA MODE. Fcp. 8vo. 1s. 6d. boards.

SWEETS AND SUPPER DISHES À LA MODE. Fcp. 8vo. 1s. 6d. boards.

De Tocqueville.—*DEMOCRACY IN AMERICA.* By ALEXIS DE TOCQUEVILLE. Translated by HENRY REEVE, C.B. 2 vols. crown 8vo. 16s.

Dickinson. — *ON RENAL AND URINARY AFFECTIONS.* By W. HOWSHIP DICKINSON, M.D. Cantab. F.R.C.P. &c. With 12 Plates and 122 Woodcuts. 3 vols. 8vo. £3. 4s. 6d.

Dixon.—*RURAL BIRD LIFE;* Essays on Ornithology, with Instructions for Preserving Objects relating to that Science. By CHARLES DIXON. With 45 Woodcuts. Crown 8vo. 5s.

Dove.—*DOMESDAY STUDIES:* being the Papers read at the Meetings of the Domesday Commemoration 1886. With a Bibliography of Domesday Book and Accounts of the MSS. and Printed Books exhibited at the Public Record Office and at the British Museum. Edited by P. EDWARD DOVE, of Lincoln's Inn, Barrister-at-Law, Honorary Secretary of the Domesday Commemoration Committee. Vol. I. 4to. 18s.; Vol. II. 4to. 18s.

Dowell.—*A HISTORY OF TAXATION AND TAXES IN ENGLAND FROM THE EARLIEST TIMES TO THE YEAR 1885.* By STEPHEN DOWELL, Assistant Solicitor of Inland Revenue. Second Edition, Revised and Altered. (4 vols. 8vo.) Vols. I. and II. The History of Taxation, 21s. Vols. III. and IV. The History of Taxes, 21s.

Doyle.—*THE OFFICIAL BARONAGE OF ENGLAND.* By JAMES E. DOYLE. Showing the Succession, Dignities, and Offices of every Peer from 1066 to 1885. Vols. I. to III. With 1,600 Portraits, Shields of Arms, Autographs, &c. 3 vols. 4to. £5. 5s.

Large-paper Edition, 3 vols. £15. 15s.

Doyle.—*WORKS BY J. A. DOYLE,* Fellow of All Souls College, Oxford.

THE ENGLISH IN AMERICA: VIRGINIA, MARYLAND, AND THE CAROLINAS. 8vo. 18s.

THE ENGLISH IN AMERICA: THE PURITAN COLONIES. 2 vols. 8vo. 36s.

Dublin University Press Series (The) : a Series of Works undertaken by the Provost and Senior Fellows of Trinity College, Dublin.

Abbott's (T. K.) Codex Rescriptus Dublinensis of St. Matthew. 4to. 21s.

——————— Evangeliorum Versio Antehieronymiana ex Codice Usseriano (Dublinensi). 2 vols. crown 8vo. 21s.

Burnside (W. S.) and Panton's (A. W.) Theory of Equations. 8vo. 12s. 6d.

Casey's (John) Sequel to Euclid's Elements. Crown 8vo. 3s. 6d.

——————— Analytical Geometry of the Conic Sections. Crown 8vo. 7s. 6d.

Davies's (J. F.) Eumenides of Æschylus. With Metrical English Translation. 8vo. 7s.

Dublin Translations into Greek and Latin Verse. Edited by R. Y. Tyrrell. 8vo. 12s. 6d.

Graves's (R. P.) Life of Sir William Hamilton. (3 vols.) Vols. I. and II. 8vo. each 15s.

Griffin (R. W.) on Parabola, Ellipse, and Hyperbola, treated Geometrically. Crown 8vo. 6s.

Haughton's (Dr. S.) Lectures on Physical Geography. 8vo. 15s.

Hobart's (W. K.) Medical Language of St. Luke. 8vo. 16s.

Leslie's (T. E. Cliffe) Essays in Political and Moral Philosophy. 8vo. 10s. 6d.

Macalister's (A.) Zoology and Morphology of Vertebrata. 8vo. 10s. 6d.

MacCullagh's (James) Mathematical and other Tracts. 8vo. 15s.

Maguire's (T.) Parmenides of Plato, Greek Text with English Introduction, Analysis, and Notes. 8vo. 7s. 6d.

Monck's (W. H. S.) Introduction to Logic. Crown 8vo. 5s.

Purser's (J. M.) Manual of Histology. Fcp. 8vo. 5s.

Roberts's (R. A.) Examples in the Analytic Geometry of Plane Curves. Fcp. 8vo. 5s.

Southey's (R.) Correspondence with Caroline Bowles. Edited by E. Dowden. 8vo. 14s.

Thornhill's (W. J.) The Æneid of Virgil, freely translated into English Blank Verse. Crown 8vo. 7s. 6d.

Tyrrell's (R. Y.) Cicero's Correspondence. Vols. I. and II. 8vo. each 12s.

——————— The Acharnians of Aristophanes, translated into English Verse. Crown 8vo. 2s. 6d.

Webb's (T. E.) Goethe's Faust, Translation and Notes. 8vo. 12s. 6d.

——————— The Veil of Isis : a Series of Essays on Idealism. 8vo. 10s. 6d.

Wilkins's (G.) The Growth of the Homeric Poems. 8vo. 6s.

Edersheim.—*WORKS BY THE REV. ALFRED EDERSHEIM, D.D.*

THE LIFE AND TIMES OF JESUS THE MESSIAH. 2 vols. 8vo. 24s.

PROPHECY AND HISTORY IN RELATION TO THE MESSIAH: the Warburton Lectures, delivered at Lincoln's Inn Chapel, 1880-1884. 8vo. 12s.

Ellicott. — *WORKS BY C. J. ELLICOTT, D.D.* Bishop of Gloucester and Bristol.

A CRITICAL AND GRAMMATICAL COMMENTARY ON ST. PAUL'S EPISTLES. 8vo.
I. CORINTHIANS. 16s.
GALATIANS. 8s. 6d.
EPHESIANS. 8s. 6d.
PASTORAL EPISTLES. 10s. 6d.
PHILIPPIANS, COLOSSIANS, and PHILEMON. 10s. 6d.
THESSALONIANS. 7s. 6d.

HISTORICAL LECTURES ON THE LIFE OF OUR LORD JESUS CHRIST. 8vo. 12s.

English Worthies. Edited by AN-DREW LANG, M.A. Fcp. 8vo. 2s. 6d. each.
DARWIN. By GRANT ALLEN.
MARLBOROUGH. By G. SAINTSBURY.
SHAFTESBURY (The First Earl). By H. D. TRAILL.
ADMIRAL BLAKE. By DAVID HANNAY.
RALEIGH. By EDMUND GOSSE.
STEELE. By AUSTIN DOBSON.
BEN JONSON. By J. A. SYMONDS.
CANNING. By FRANK H. HILL.
CLAVERHOUSE. By MOWBRAY MORRIS.

Epochs of Ancient History. 10 vols. fcp. 8vo. 2s. 6d. each. See p. 24.

Epochs of Church History. 8 vols. fcp. 8vo. 2s. 6d. each. See p. 24.

Epochs of Modern History. 18 vols. fcp. 8vo. 2s. 6d. each. See p. 24.

Erichsen.—*WORKS BY JOHN ERIC ERICHSEN, F.R.S.*

THE SCIENCE AND ART OF SUR-GERY: Being a Treatise on Surgical Injuries, Diseases, and Operations. With 984 Illustrations. 2 vols. 8vo. 42s.

ON CONCUSSION OF THE SPINE, NER-VOUS SHOCKS, and other Obscure Injuries of the Nervous System. Cr. 8vo. 10s. 6d.

Ewald. — *WORKS BY PROFESSOR HEINRICH EWALD,* of Göttingen.

THE ANTIQUITIES OF ISRAEL. Translated from the German by H. S. SOLLY, M.A. 8vo. 12s. 6d.

THE HISTORY OF ISRAEL. Translated from the German. 8 vols. 8vo. Vols. I. and II. 24s. Vols. III. and IV. 21s. Vol. V. 18s. Vol. VI. 16s. Vol. VII. 21s. Vol. VIII. with Index to the Complete Work. 18s.

Fairbairn.—*WORKS BY SIR W. FAIRBAIRN, BART. C.E.*

A TREATISE ON MILLS AND MILL-WORK, with 18 Plates and 333 Woodcuts. 1 vol. 8vo. 25s.

USEFUL INFORMATION FOR ENGI-NEERS. With many Plates and Woodcuts. 3 vols. crown 8vo. 31s. 6d.

Farrar. — *LANGUAGE AND LAN-GUAGES.* A Revised Edition of *Chapters on Language and Families of Speech.* By F. W. FARRAR, D.D. Crown 8vo. 6s.

Firbank.—*THE LIFE AND WORK OF JOSEPH FIRBANK, J.P. D.L.* Railway Contractor. By FREDERICK MCDERMOTT, Barrister-at-Law. 8vo. 5s.

Fitzwygram. — *HORSES AND STABLES.* By Major-General Sir F. FITZWYGRAM, Bart. With 19 pages of Illustrations. 8vo. 5s.

Forbes.—*A COURSE OF LECTURES ON ELECTRICITY,* delivered before the Society of Arts. By GEORGE FORBES, M.A. F.R.S. (L. & E.) With 17 Illustrations. Crown 8vo. 5s.

Ford.—*THE THEORY AND PRACTICE OF ARCHERY.* By the late HORACE FORD. New Edition, thoroughly Revised and Re-written by W. BUTT, M.A. With a Preface by C. J. LONGMAN, M.A. F.S.A. 8vo. 14s.

Fox.—*THE EARLY HISTORY OF CHARLES JAMES FOX.* By the Right Hon. Sir G. O. TREVELYAN, Bart.
Library Edition, 8vo. 18s.
Cabinet Edition, cr. 8vo. 6s.

Francis.—*A BOOK ON ANGLING;* or, Treatise on the Art of Fishing in every branch; including full Illustrated Lists of Salmon Flies. By FRANCIS FRANCIS. Post 8vo. Portrait and Plates, 15s.

Freeman.—*THE HISTORICAL GEO-GRAPHY OF EUROPE.* By E. A. FREEMAN, D.C.L. With 65 Maps. 2 vols. 8vo. 31s. 6d.

Froude.—*WORKS BY JAMES A. FROUDE, M.A.*

THE HISTORY OF ENGLAND, from the Fall of Wolsey to the Defeat of the Spanish Armada.
Cabinet Edition, 12 vols. cr. 8vo. £3. 12s.
Popular Edition, 12 vols. cr. 8vo. £2. 2s.

SHORT STUDIES ON GREAT SUBJECTS. 4 vols. crown 8vo. 24s.

CÆSAR : a Sketch. Crown 8vo. 6s.

THE ENGLISH IN IRELAND IN THE EIGHTEENTH CENTURY. 3 vols. crown 8vo. 18s.

OCEANA ; OR, ENGLAND AND HER COLONIES. With 9 Illustrations. Crown 8vo. 2s. boards, 2s. 6d. cloth.

THE ENGLISH IN THE WEST INDIES; OR, THE BOW OF ULYSSES. With 9 Illustrations. 8vo. 18s.

THOMAS CARLYLE, a History of the first Forty Years of his Life, 1795 to 1835. 2 vols. 8vo. 32s.

THOMAS CARLYLE, a History of His Life in London from 1834 to his death in 1881. With Portrait engraved on steel. 2 vols. 8vo. 32s.

Galloway. — *THE FUNDAMENTAL PRINCIPLES OF CHEMISTRY PRACTICALLY TAUGHT BY A NEW METHOD.* By ROBERT GALLOWAY, M.R.I.A. F.C.S. Crown 8vo. 6s. 6d.

Ganot. — *WORKS BY PROFESSOR GANOT.* Translated by E. ATKINSON, Ph.D. F.C.S.

ELEMENTARY TREATISE ON PHYSICS. With 5 Coloured Plates and 923 Woodcuts. Crown 8vo. 15s.

NATURAL PHILOSOPHY FOR GENERAL READERS AND YOUNG PERSONS. With 2 Plates, 518 Woodcuts, and an Appendix of Questions. Cr. 8vo. 7s. 6d.

Gardiner. — *WORKS BY SAMUEL RAWSON GARDINER, LL.D.*

HISTORY OF ENGLAND, from the Accession of James I. to the Outbreak of the Civil War, 1603-1642. Cabinet Edition, thoroughly revised. 10 vols. crown 8vo. price 6s. each.

A HISTORY OF THE GREAT CIVIL WAR, 1642-1649. (3 vols.) Vol. I. 1642-1644. With 24 Maps. 8vo. 21s.
[*Continued above.*]

Gardiner.—*WORKS BY S. R. GARDINER, LL.D.—continued.*

OUTLINE OF ENGLISH HISTORY, B.C. 55-A.D. 1886. With 96 Woodcuts, fcp. 8vo. 2s. 6d.

** For other Works, see ' Epochs of Modern History,' p. 24.

Garrod.—*WORKS BY SIR ALFRED BARING GARROD, M.D. F.R.S.*

A TREATISE ON GOUT AND RHEUMATIC GOUT (RHEUMATOID ARTHRITIS). With 6 Plates, comprising 21 Figures (14 Coloured), and 27 Illustrations engraved on Wood. 8vo. 21s.

THE ESSENTIALS OF MATERIA MEDICA AND THERAPEUTICS. New Edition, revised and adapted to the New Edition of the British Pharmacopœia, by NESTOR TIRARD, M.D. Crown 8vo. 12s. 6d.

Gilkes.—*BOYS AND MASTERS :* a Story of School Life. By A. H. GILKES, M.A. Head Master of Dulwich College. Crown 8vo. 3s. 6d.

Goethe.—*FAUST.* A New Translation, chiefly in Blank Verse ; with Introduction and Notes. By JAMES ADEY BIRDS, B.A. F.G.S. Crown 8vo. 12s. 6d.

FAUST. The German Text, with an English Introduction and Notes for Students. By ALBERT M. SELSS, M.A. Ph.D. Crown 8vo. 5s.

Goodeve.—*WORKS BY T. M. GOODEVE, M.A.*

PRINCIPLES OF MECHANICS. With 253 Woodcuts. Crown 8vo. 6s.

THE ELEMENTS OF MECHANISM. With 342 Woodcuts. Crown 8vo. 6s.

A MANUAL OF MECHANICS : an Elementary Text-Book for Students of Applied Mechanics. With 138 Illustrations and Diagrams, and 141 Examples. Fcp. 8vo. 2s. 6d.

Grant.—*THE ETHICS OF ARISTOTLE.* The Greek Text illustrated by Essays and Notes. By Sir ALEXANDER GRANT, Bart. LL.D. D.C.L. &c. 2 vols. 8vo. 32s.

Gray. — *ANATOMY, DESCRIPTIVE AND SURGICAL.* By HENRY GRAY, F.R.S. late Lecturer on Anatomy at St. George's Hospital. With 569 Woodcut Illustrations, a large number of which are coloured. Re-edited by T. PICKERING PICK, Surgeon to St. George's Hospital. Royal 8vo. 36s.

Green.—THE WORKS OF THOMAS HILL GREEN, late Fellow of Balliol College, and Whyte's Professor of Moral Philosophy in the University of Oxford. Edited by R. L. NETTLESHIP, Fellow of Balliol College, Oxford (3 vols.) Vols. I. and II.—Philosophical Works. 8vo. 16s. each.

Greville.—A JOURNAL OF THE REIGNS OF KING GEORGE IV. KING WILLIAM IV. AND QUEEN VICTORIA. By the late CHARLES C. F. GREVILLE, Esq. Clerk of the Council to those Sovereigns. Edited by HENRY REEVE, C.B. D.C.L. Corresponding Member of the Institute of France. 8 vols. Crown 8vo. 6s. each. (*In course of Publication in Monthly Volumes.*)

Grove.—THE CORRELATION OF PHYSICAL FORCES. By the Hon. Sir W. R. GROVE, F.R.S. &c. 8vo. 15s.

Gwilt.—AN ENCYCLOPÆDIA OF ARCHITECTURE. By JOSEPH GWILT, F.S.A. Illustrated with more than 1,100 Engravings on Wood. Revised, with Alterations and Considerable Additions, by WYATT PAPWORTH. 8vo. 52s. 6d.

Haggard.—WORKS BY H. RIDER HAGGARD.

SHE : A HISTORY OF ADVENTURE. Crown 8vo. 6s.

ALLAN QUATERMAIN. With 31 Illustrations by C. H. M. KERR. Crown 8vo. 6s.

Halliwell-Phillipps.—OUTLINES OF THE LIFE OF SHAKESPEARE. By J. O. HALLIWELL-PHILLIPPS, F.R.S. 2 vols. Royal 8vo. 10s. 6d.

Harte.—NOVELS BY BRET HARTE.

IN THE CARQUINEZ WOODS. Fcp. 8vo. 1s. boards; 1s. 6d. cloth.

ON THE FRONTIER. Three Stories. 16mo. 1s.

BY SHORE AND SEDGE. Three Stories. 16mo. 1s.

Hartwig.—WORKS BY DR. G. HARTWIG.

THE SEA AND ITS LIVING WONDERS. With 12 Plates and 303 Woodcuts. 8vo. 10s. 6d.

THE TROPICAL WORLD. With 8 Plates, and 172 Woodcuts. 8vo. 10s. 6d.

THE POLAR WORLD. With 3 Maps, 8 Plates, and 85 Woodcuts. 8vo. 10s. 6d.
[*Continued above.*

Hartwig. — WORKS BY DR. G. HARTWIG.—*continued.*

THE SUBTERRANEAN WORLD. With 3 Maps and 80 Woodcuts. 8vo. 10s. 6d.

THE AERIAL WORLD. With Map, 8 Plates, and 60 Woodcuts. 8vo. 10s. 6d.

The following books are extracted from the foregoing works by Dr. HARTWIG :—

HEROES OF THE ARCTIC REGIONS. With 19 Illustrations. Crown 8vo. 2s. cloth extra, gilt edges.

WONDERS OF THE TROPICAL FORESTS. With 40 Illustrations. Crown 8vo. 2s. cloth extra, gilt edges.

WORKERS UNDER THE GROUND; or, Mines and Mining. With 29 Illustrations. Crown 8vo. 2s. cloth extra, gilt edges.

MARVELS OVER OUR HEADS. With 29 Illustrations. Crown 8vo. 2s. cloth extra, gilt edges.

MARVELS UNDER OUR FEET. With 22 Illustrations. Crown 8vo. 2s. cloth extra, gilt edges.

DWELLERS IN THE ARCTIC REGIONS. With 29 Illustrations. Crown 8vo. 2s. 6d. cloth extra, gilt edges.

WINGED LIFE IN THE TROPICS. With 55 Illustrations. Crown 8vo. 2s. 6d. cloth extra, gilt edges.

VOLCANOES AND EARTHQUAKES. With 30 Illustrations. Crown 8vo. 2s. 6d. cloth extra, gilt edges.

WILD ANIMALS OF THE TROPICS. With 66 Illustrations. Crown 8vo. 3s. 6d. cloth extra, gilt edges.

SEA MONSTERS AND SEA BIRDS. With 75 Illustrations. Crown 8vo. 2s. 6d. cloth extra, gilt edges.

DENIZENS OF THE DEEP. With 117 Illustrations. Crown 8vo. 2s. 6d. cloth extra, gilt edges.

Hassall.—THE INHALATION TREATMENT OF DISEASES OF THE ORGANS OF RESPIRATION, including Consumption. By ARTHUR HILL HASSALL, M.D. With 19 Illustrations of Apparatus. Cr. 8vo. 12s. 6d.

Havelock. — MEMOIRS OF SIR HENRY HAVELOCK, K.C.B. By JOHN CLARK MARSHMAN. Crown 8vo. 3s. 6d.

Hearn.—THE GOVERNMENT OF ENGLAND; its Structure and its Development. By WILLIAM EDWARD HEARN, Q.C. 8vo. 16s.

Helmholtz.—*WORKS BY PROFESSOR HELMHOLTZ.*

ON THE SENSATIONS OF TONE AS A PHYSIOLOGICAL BASIS FOR THE THEORY OF MUSIC. Royal 8vo. 28s.

POPULAR LECTURES ON SCIENTIFIC SUBJECTS. With 68 Woodcuts. 2 vols. Crown 8vo. 15s. or separately, 7s. 6d. each.

Herschel.—*OUTLINES OF ASTRONOMY.* By Sir J. F. W. HERSCHEL, Bart. M.A. With Plates and Diagrams. Square crown 8vo. 12s.

Hester's Venture: a Novel. By the Author of 'The Atelier du Lys.' Crown 8vo. 2s. 6d.

Hewitt. — *THE DIAGNOSIS AND TREATMENT OF DISEASES OF WOMEN, INCLUDING THE DIAGNOSIS OF PREGNANCY.* By GRAILY HEWITT, M.D. With 211 Engravings. 8vo. 24s.

Historic Towns. Edited by E. A. FREEMAN, D.C.L. and Rev. WILLIAM HUNT, M.A. With Maps and Plans. Crown 8vo. 3s. 6d. each.

LONDON. By W. E. LOFTIE.

EXETER. By E. A. FREEMAN.

BRISTOL. By W. HUNT.

OXFORD. By C. W. BOASE.

*** Other Volumes are in preparation.

Hobart.—*SKETCHES FROM MY LIFE.* By Admiral HOBART PASHA. With Portrait. Crown 8vo. 7s. 6a.

Holmes.—*A SYSTEM OF SURGERY,* Theoretical and Practical, in Treatises by various Authors. Edited by TIMOTHY HOLMES, M.A. and J. W. HULKE, F.R.S. 3 vols. royal 8vo. £4. 4s.

Homer.—*THE ILIAD OF HOMER,* Homometrically translated by C. B. CAYLEY. 8vo. 12s. 6d.

THE ILIAD OF HOMER. The Greek Text, with a Verse Translation, by W. C. GREEN, M.A. Vol. I. Books I.-XII. Crown 8vo. 6s.

Hopkins.—*CHRIST THE CONSOLER;* a Book of Comfort for the Sick. By ELLICE HOPKINS. Fcp. 8vo. 2s. 6d.

Howitt.—*VISITS TO REMARKABLE PLACES,* Old Halls, Battle-Fields, Scenes illustrative of Striking Passages in English History and Poetry. By WILLIAM HOWITT. With 80 Illustrations engraved on Wood. Crown 8vo. 5s.

Hudson & Gosse.—*THE ROTIFE[R]* OR 'WHEEL-ANIMALCULES.' By C. HUDSON, LL.D. and P. H. GOSS[E] F.R.S. With 30 Coloured Plates. In Parts. 4to. 10s. 6d. each. Complete 2 vols. 4to. £3. 10s.

Hullah.—*WORKS BY JOHN HU[L]LAH, LL.D.*

COURSE OF LECTURES ON THE H[IS]TORY OF MODERN MUSIC. 8vo. 8s.

COURSE OF LECTURES ON THE TRA[N]SITION PERIOD OF MUSICAL HISTO[RY] 8vo. 10s. 6d.

Hume.—*THE PHILOSOPHICAL WOR[KS]* OF DAVID HUME. Edited by T. GREEN, M.A. and the Rev. T. GROSE, M.A. 4 vols. 8vo. 56s. separately, Essays, 2 vols. 28s. Trea[tise] of Human Nature. 2 vols. 28s.

Huth.—*THE MARRIAGE OF NE[AR] KIN,* considered with respect to the L[aws] of Nations, the Result of Experien[ce] and the Teachings of Biology. ALFRED H. HUTH. Royal 8vo. 21s.

In the Olden Time: a Tale [of] the Peasant War in Germany. By [the] Author of 'Mademoiselle Mori.' Cro[wn] 8vo. 2s. 6d.

Ingelow.—*WORKS BY JEAN INGE[LOW.]*

POETICAL WORKS. Vols. 1 and [2.] Fcp. 8vo. 12s.

LYRICAL AND OTHER POEMS. S[e]lected from the Writings of J[EAN] INGELOW. Fcp. 8vo. 2s. 6d. cloth plai[n.] 3s. cloth gilt.

Jackson.—*AID TO ENGINEERI[NG] SOLUTION.* By LOWIS D'A. JACKS[ON,] C.E. With 111 Diagrams and 5 W[ood]cut Illustrations. 8vo. 21s.

James.—*THE LONG WHITE MOU[N]TAIN;* or, a Journey in Manchuria, w[ith] an Account of the History, Administ[ra]tion, and Religion of that Province. [By] H. E. JAMES, of Her Majesty's Bom[bay] Civil Service. With Illustrations and [a] Map. 1 vol. 8vo. 24s.

Jameson.—*WORKS BY MRS. JAM[E]SON.*

LEGENDS OF THE SAINTS AND MA[R]TYRS. With 19 Etchings and 187 W[ood]cuts. 2 vols. 31s. 6d.

LEGENDS OF THE MADONNA, t[he] Virgin Mary as represented in Sac[red] and Legendary Art. With 27 Etchin[gs] and 165 Woodcuts. 1 vol. 21s.

[Continued on next page.

Jameson.—*WORKS BY MRS. JAME-SON—continued.*

LEGENDS OF THE MONASTIC ORDERS. With 11 Etchings and 88 Woodcuts. 1 vol. 21*s.*

HISTORY OF THE SAVIOUR, His Types and Precursors. Completed by Lady EASTLAKE. With 13 Etchings and 281 Woodcuts. 2 vols. 42*s.*

Jeans.—*WORKS BY J. S. JEANS.*

ENGLAND'S SUPREMACY: its Sources, Economics, and Dangers. 8vo. 8*s.* 6*d.*

RAILWAY PROBLEMS: An Inquiry into the Economic Conditions of Railway Working in Different Countries. 8vo. 12*s.* 6*d.*

Jenkin. — *PAPERS, LITERARY, SCIENTIFIC, &c.* By the late FLEEMING JENKIN, F.R.S.S. L. & E. Professor of Engineering in the University of Edinburgh. Edited by SIDNEY COLVIN, M.A. and J. A. EWING, F.R.S. With Memoir by ROBERT LOUIS STEVENSON, and Facsimiles of Drawings by Fleeming Jenkin. 2 vols. 8vo. 32*s.*

Johnson.—*THE PATENTEE'S MAN-UAL;* a Treatise on the Law and Practice of Letters Patent. By J. JOHNSON and J. H. JOHNSON. 8vo. 10*s.* 6*d.*

Johnston.—*A GENERAL DICTION-ARY OF GEOGRAPHY*, Descriptive, Physical, Statistical, and Historical; a complete Gazetteer of the World. By KEITH JOHNSTON. Medium 8vo. 42*s.*

Johnstone.—*A SHORT INTRODUC-TION TO THE STUDY OF LOGIC.* By LAURENCE JOHNSTONE. Crown 8vo. 2*s.* 6*d.*

Jordan. — *WORKS BY WILLIAM LEIGHTON JORDAN, F.R.G.S.*

THE OCEAN: a Treatise on Ocean Currents and Tides and their Causes. 8vo. 21*s.*

THE NEW PRINCIPLES OF NATURAL PHILOSOPHY. With 13 plates. 8vo. 21*s.*

THE WINDS: an Essay in Illustration of the New Principles of Natural Philosophy. Crown 8vo. 2*s.*

THE STANDARD OF VALUE. Crown 8vo. 5*s.*

Jukes.—*WORKS BY ANDREW JUKES.*

THE NEW MAN AND THE ETERNAL LIFE. Crown 8vo. 6*s.*

THE TYPES OF GENESIS. Crown 8vo. 7*s.* 6*d.*

THE SECOND DEATH AND THE RE-STITUTION OF ALL THINGS. Crown 8vo. 3*s.* 6*d.*

THE MYSTERY OF THE KINGDOM. Crown 8vo. 2*s.* 6*d.*

Justinian. — *THE INSTITUTES OF JUSTINIAN;* Latin Text, chiefly that of Huschke, with English Introduction, Translation, Notes, and Summary. By THOMAS C. SANDARS, M.A. 8vo. 18*s.*

Kalisch. — *WORKS BY M. M. KALISCH, M.A.*

BIBLE STUDIES. Part I. The Prophecies of Balaam. 8vo. 10*s.* 6*d.* Part II. The Book of Jonah. 8vo. 10*s.* 6*d.*

COMMENTARY ON THE OLD TESTA-MENT; with a New Translation. Vol. I. Genesis, 8vo. 18*s.* or adapted for the General Reader, 12*s.* Vol. II. Exodus, 15*s.* or adapted for the General Reader, 12*s.* Vol. III. Leviticus, Part I. 15*s.* or adapted for the General Reader, 8*s.* Vol. IV. Leviticus, Part II. 15*s.* or adapted for the General Reader, 8*s.*

HEBREW GRAMMAR. With Exercises. Part I. 8vo. 12*s.* 6*d.* Key, 5*s.* Part II. 12*s.* 6*d.*

Kant.—*WORKS BY EMMANUEL KANT.*

CRITIQUE OF PRACTICAL REASON. Translated by Thomas Kingsmill Abbott, B.D. 8vo. 12*s.* 6*d.*

INTRODUCTION TO LOGIC, AND HIS ESSAY ON THE MISTAKEN SUBTILTY OF THE FOUR FIGURES. Translated by Thomas Kingsmill Abbott, B.D. With a few Notes by S. T. Coleridge. 8vo. 6*s.*

Kendall.—*WORKS BY MAY KEN-DALL.*

FROM A GARRET. Crown 8vo. 6*s.*

DREAMS TO SELL; Poems. Fcp. 8vo. 6*s.*

Killick.—*HANDBOOK TO MILL'S SYSTEM OF LOGIC.* By the Rev. A. H. KILLICK, M.A. Crown 8vo. 3*s.* 6*d.*

Kirkup.—*AN INQUIRY INTO SOCIAL-ISM.* By THOMAS KIRKUP, Author of the Article on 'Socialism' in the 'Encyclopædia Britannica.' Crown 8vo. 5*s.*

Knowledge Library. (*See* PROCTOR'S Works, p. 17.)

Kolbe.—*A Short Text-book of Inorganic Chemistry.* By Dr. Hermann Kolbe. Translated from the German by T. S. Humpidge, Ph.D. With a Coloured Table of Spectra and 66 Illustrations. Crown 8vo. 7s. 6d.

Ladd. — *Elements of Physiological Psychology:* a Treatise of the Activities and Nature of the Mind from the Physical and Experimental Point of View. By George T. Ladd. With 113 Illustrations and Diagrams. 8vo. 21s.

Lang.—*Works by Andrew Lang.*

Myth, Ritual, and Religion. 2 vols. crown 8vo. 21s.

Custom and Myth; Studies of Early Usage and Belief. With 15 Illustrations. Crown 8vo. 7s. 6d.

Letters to Dead Authors. Fcp. 8vo. 6s. 6d.

Books and Bookmen. With 2 Coloured Plates and 17 Illustrations. Cr. 8vo. 6s. 6d.

Johnny Nut and the Golden Goose. Done into English by Andrew Lang, from the French of Charles Deulin. Illustrated by Am. Lynen. Royal 8vo. 10s. 6d. gilt edges.

Ballads of Books. Edited by Andrew Lang. Fcp. 8vo. 6s.

Larden.—*Electricity for Public Schools and Colleges.* With numerous Questions and Examples with Answers, and 214 Illustrations and Diagrams. By W. Larden, M.A. Crown 8vo. 6s.

Laughton.—*Studies in Naval History;* Biographies. By J. K. Laughton, M.A. Professor of Modern History at King's College, London. 8vo. 10s. 6d.

Lecky.—*Works by W. E. H. Lecky.*

History of England in the Eighteenth Century. 8vo. Vols. I. & II. 1700–1760. 36s. Vols. III. & IV. 1760–1784. 36s. Vols. V. & VI. 1784–1793. 36s.

The History of European Morals from Augustus to Charlemagne. 2 vols. crown 8vo. 16s.

History of the Rise and Influence of the Spirit of Rationalism in Europe. 2 vols. crown 8vo. 16s.

Lewes.—*The History of Philosophy,* from Thales to Comte. By George Henry Lewes. 2 vols. 8vo. 32s.

Lindt.—*Picturesque New Guine* By J. W. Lindt, F.R.G.S. With Full-page Photographic Illustrations produced by the Autotype Compan Crown 4to. 42s.

Liveing.—*Works by Robert Liv ing, M.A. and M.D. Cantab.*

Handbook on Diseases of th Skin. With especial reference to D nosis and Treatment. Fcp 8vo. 5s.

Notes on the Treatment of Ski Diseases. 18mo. 3s.

Lloyd.—*A Treatise on Magne ism,* General and Terrestrial. By Lloyd, D.D. D.C.L. 8vo. 10s. 6d.

Lloyd.—*The Science of Agricu ture.* By F. J. Lloyd. 8vo. 12s.

Longman.—*History of the Li and Times of Edward III.* William Longman, F.S.A. W 9 Maps, 8 Plates, and 16 Woodcuts. vols. 8vo. 28s.

Longman.— *Works by Frederi W. Longman, Balliol College, O*

Chess Openings. Fcp. 8vo. 2s.

Frederick the Great and Seven Years' War. With 2 Col Maps. 8vo. 2s. 6d.

A New Pocket Dictionary the German and English guages. Square 18mo. 2s. 6d.

Longman's Magazine. Publish Monthly. Price Sixpence. Vols. 1–10, 8vo. price 5s. each.

Longmore.— *Gunshot Injuri* Their History, Characteristic Fea Complications, and General Treatm By Surgeon-General Sir T. Longmo C.B., F.R.C.S. With 58 Illustratio 8vo. 31s. 6d.

Loudon.—*Works by J. C. Lou F.L.S.*

Encyclopædia of Gardeni the Theory and Practice of Horticul Floriculture, Arboriculture, and scape Gardening. With 1,000 Wood 8vo. 21s.

Encyclopædia of Agricultu the Laying-out, Improvement, Management of Landed Property ; Cultivation and Economy of the P tions of Agriculture. With 1,100 W cuts. 8vo. 21s.

Encyclopædia of Plants ; Specific Character, Description, Cult History, &c. of all Plants found in G Britain. With 12,000 Woodcuts. 8vo.

Lubbock.—*THE ORIGIN OF CIVILIZATION AND THE PRIMITIVE CONDITION OF MAN.* By Sir J. LUBBOCK, Bart. M.P. F.R.S. With Illustrations. 8vo. 18s.

Lyall.—*THE AUTOBIOGRAPHY OF A SLANDER.* By EDNA LYALL, Author of 'Donovan,' 'We Two,' &c. Fcp. 8vo. 1s. sewed.

Lyra Germanica ; Hymns Translated from the German by Miss C. WINKWORTH. Fcp. 8vo. 5s.

Macaulay.—*WORKS AND LIFE OF LORD MACAULAY.*

HISTORY OF ENGLAND FROM THE ACCESSION OF JAMES THE SECOND:
Student's Edition, 2 vols. crown 8vo. 12s.
People's Edition, 4 vols. crown 8vo. 16s.
Cabinet Edition, 8 vols. post 8vo. 48s.
Library Edition, 5 vols. 8vo. £4.

CRITICAL AND HISTORICAL ESSAYS, with *LAYS of ANCIENT ROME*, in 1 volume :
Authorised Edition, crown 8vo. 2s. 6d. or 3s. 6d. gilt edges.
Popular Edition, crown 8vo. 2s. 6d.

CRITICAL AND HISTORICAL ESSAYS:
Student's Edition, 1 vol. crown 8vo. 6s.
People's Edition, 2 vols. crown 8vo. 8s.
Cabinet Edition, 4 vols. post 8vo. 24s.
Library Edition, 3 vols. 8vo. 36s.

ESSAYS which may be had separately price 6d. each sewed, 1s. each cloth :
Addison and Walpole.
Frederick the Great.
Croker's Boswell's Johnson.
Hallam's Constitutional History.
Warren Hastings. (3d. sewed, 6d. cloth.)
The Earl of Chatham (Two Essays).
Ranke and Gladstone.
Milton and Machiavelli.
Lord Bacon.
Lord Clive.
Lord Byron, and The Comic Dramatists of the Restoration.

The Essay on Warren Hastings annotated by S. HALES, 1s. 6d.
The Essay on Lord Clive annotated by H. COURTHOPE BOWEN, M.A. 2s. 6d.

SPEECHES :
People's Edition, crown 8vo. 3s. 6d.

MISCELLANEOUS WRITINGS :
Library Edition, 2 vols. 8vo. 21s.
People's Edition, 1 vol. crown 8vo. 4s. 6d.
[*Continued above.*

Macaulay—*WORKS AND LIFE OF LORD MACAULAY*—*continued.*
LAYS OF ANCIENT ROME, &c.
Illustrated by G. Scharf, fcp. 4to. 10s. 6d.
————————— Popular Edition, fcp. 4to. 6d. sewed, 1s. cloth.
Illustrated by J. R. Weguelin, crown 8vo. 3s. 6d. cloth extra, gilt edges.
Cabinet Edition, post 8vo. 3s. 6d.
Annotated Edition, fcp. 8vo. 1s. sewed 1s. 6d. cloth, or 2s. 6d. cloth extra, gilt edges.
SELECTIONS FROM THE WRITINGS OF LORD MACAULAY. Edited, with Occasional Notes, by the Right Hon. Sir G. O. TREVELYAN, Bart. Crown 8vo. 6s.
MISCELLANEOUS WRITINGS AND SPEECHES :
Student's Edition, in ONE VOLUME, crown 8vo. 6s.
Cabinet Edition, including Indian Penal Code, Lays of Ancient Rome, and Miscellaneous Poems, 4 vols. post 8vo. 24s.
THE COMPLETE WORKS OF LORD MACAULAY. Edited by his Sister, Lady TREVELYAN.
Library Edition, with Portrait, 8 vols. demy 8vo. £5. 5s.
Cabinet Edition, 16 vols. post 8vo. £4. 16s.
THE LIFE AND LETTERS OF LORD MACAULAY. By the Right Hon. Sir G. O. TREVELYAN, Bart.
Popular Edition, 1 vol. crown 8vo.
Cabinet Edition, 2 vols. post 8vo.
Library Edition, 2 vols. 8vo. 36s.

Macdonald.—*WORKS BY GEORGE MACDONALD, LL.D.*
UNSPOKEN SERMONS. First Series. Crown 8vo. 3s. 6d.
UNSPOKEN SERMONS. Second Series. Crown 8vo. 3s. 6d.
THE MIRACLES OF OUR LORD. Crown 8vo. 3s. 6d.
A BOOK OF STRIFE, IN THE FORM OF THE DIARY OF AN OLD SOUL: Poems. 12mo. 6s.

Macfarren.—*WORKS BY SIR G. A. MACFARREN.*
LECTURES ON HARMONY. 8vo. 12s.
ADDRESSES AND LECTURES. Crown 8vo. 6s. 6d.

Macleod.—*WORKS BY HENRY D. MACLEOD, M.A.*
THE ELEMENTS OF ECONOMICS. In 2 vols. Vol. I. crown 8vo. 7s. 6d. Vol. II. PART I, crown 8vo. 7s. 6d.
THE ELEMENTS OF BANKING. Crown 8vo. 5s.
THE THEORY AND PRACTICE OF BANKING. Vol. I. 8vo. 12s. Vol. II. 14s.

McCulloch. — *The Dictionary of Commerce and Commercial Navigation* of the late J. R. McCulloch, of H.M. Stationery Office. Latest Edition, containing the most recent Statistical Information by A. J. Wilson. 1 vol. medium 8vo. with 11 Maps and 30 Charts, price 63*s*. cloth, or 70*s*. strongly half-bound in russia.

Mademoiselle Mori: a Tale of Modern Rome. By the Author of 'The Atelier du Lys.' Crown 8vo. 2*s*. 6*d*.

Mahaffy. — *A History of Classical Greek Literature.* By the Rev. J. P. Mahaffy, M.A. Crown 8vo. Vol. I. Poets, 7*s*. 6*d*. Vol. II. Prose Writers, 7*s*. 6*d*.

Malmesbury. — *Memoirs of an Ex-Minister:* an Autobiography. By the Earl of Malmesbury, G.C.B. Crown 8vo. 7*s*. 6*d*.

Manning. — *The Temporal Mission of the Holy Ghost;* or, Reason and Revelation. By H. E. Manning, D.D. Cardinal-Archbishop. Crown 8vo. 8*s*. 6*d*.

Martin. — *Navigation and Nautical Astronomy.* Compiled by Staff-Commander W. R. Martin, R.N. Instructor in Surveying, Navigation, and Compass Adjustment; Lecturer on Meteorology at the Royal Naval College, Greenwich. Sanctioned for use in the Royal Navy by the Lords Commissioners of the Admiralty. Royal 8vo. 18*s*.

Martineau — *Works by James Martineau, D.D.*
Hours of Thought on Sacred Things. Two Volumes of Sermons. 2 vols. crown 8vo. 7*s*. 6*d*. each.
Endeavours after the Christian Life. Discourses. Crown 8vo. 7*s*. 6*d*.

Maunder's Treasuries.
Biographical Treasury. Reconstructed, revised, and brought down to the year 1882, by W. L. R. Cates. Fcp. 8vo. 6*s*.
Treasury of Natural History; or, Popular Dictionary of Zoology. Fcp. 8vo. with 900 Woodcuts, 6*s*.
Treasury of Geography, Physical, Historical, Descriptive, and Political. With 7 Maps and 16 Plates. Fcp. 8vo. 6*s*.
Historical Treasury: Outlines of Universal History, Separate Histories of all Nations. Revised by the Rev. Sir G. W. Cox, Bart. M.A. Fcp. 8vo. 6*s*.
[Continued above.

Maunder's Treasuries — *cont*
Treasury of Knowledg Library of Reference. C an English Dictionary and Universal Gazetteer, Classical Dictiona Chronology, Law Dictionary, &c.] 8vo. 6*s*.
Scientific and Literary Tr sury: a Popular Encyclopædia of Scien Literature, and Art. Fcp. 8vo. 6*s*.
The Treasury of Bible Kno ledge; being a Dictionary of the Boo Persons, Places, Events, and other matt of which mention is made in Holy Scr ture. By the Rev. J. Ayre, M.A. W 5 Maps, 15 Plates, and 300 Woodcu Fcp. 8vo. 6*s*.
The Treasury of Botany, Popular Dictionary of the Vegeta Kingdom. Edited by J. Lindley, F.R. and T. Moore, F.L.S. With 274 W cuts and 20 Steel Plates. Two P fcp. 8vo. 12*s*.

Max Müller. — *Works by F. M Müller, M.A.*
Biographical Essays. Crown 8v 7*s*. 6*d*.
Selected Essays on Languag Mythology and Religion. 2 vo crown 8vo. 16*s*.
Lectures on the Science of La guage. 2 vols. crown 8vo. 16*s*.
India, What Can it Teach Us A Course of Lectures delivered before t University of Cambridge. 8vo. 12*s*.
Hibbert Lectures on the Orig and Growth of Religion, as ill trated by the Religions of India. Crow 8vo. 7*s*. 6*d*.
Introduction to the Science o Religion: Four Lectures delivered at th Royal Institution. Crown 8vo. 7*s*. 6*d*.
The Science of Thought. 8vo. 21*s*
Biographies of Words, and th Home of the Aryas. Crown 8vo. 7*s*. 6*d*.
A Sanskrit Grammar for Be ginners. New and Abridged Edition, accented and transliterated throughout, with a chapter on Syntax and an Appendix on Classical Metres. By A. A. MacDonell, M.A. Ph.D. Crown 8vo. 6*s*.

May. — *Works by the Right Hon. Sir Thomas Erskine May, K.C.B.*
The Constitutional History of England since the Accession of George III. 1760–1870. 3 vols. crown 8vo. 18*s*.
Democracy in Europe; a History. 2 vols. 8vo. 32*s*.

Meath.—*WORKS BY THE EARL OF MEATH (Lord Brabazon).*

SOCIAL ARROWS: Reprinted Articles on various Social Subjects. Crown 8vo. 1s. boards, 5s. cloth.

PROSPERITY OR PAUPERISM? Physical, Industrial, and Technical Training. (Edited by the EARL OF MEATH). 8vo. 5s.

Melville.—*NOVELS BY G. J. WHYTE MELVILLE.* Crown 8vo. 1s. each, boards; 1s. 6d. each, cloth.

The Gladiators.	Holmby House.
The Interpreter.	Kate Coventry.
Good for Nothing.	Digby Grand.
The Queen's Maries.	General Bounce.

Mendelssohn.—*THE LETTERS OF FELIX MENDELSSOHN.* Translated by Lady WALLACE. 2 vols. crown 8vo. 10s.

Merivale.—*WORKS BY THE VERY REV. CHARLES MERIVALE, D.D. Dean of Ely.*

HISTORY OF THE ROMANS UNDER THE EMPIRE. 8 vols. post 8vo. 48s.

THE FALL OF THE ROMAN REPUBLIC: a Short History of the Last Century tury of the Commonwealth. 12mo. 7s. 6d.

GENERAL HISTORY OF ROME FROM B.C. 753 TO A.D. 476. Crown 8vo. 7s. 6d.

THE ROMAN TRIUMVIRATES. With Maps. Fcp. 8vo. 2s. 6d.

Meyer.—*MODERN THEORIES OF CHEMISTRY.* By Professor LOTHAR MEYER. Translated, from the Fifth Edition of the German, by P. PHILLIPS BEDSON, D.Sc. (Lond.) B.Sc. (Vict.) F.C.S.; and W. CARLETON WILLIAMS, B.Sc. (Vict.) F.C.S. 8vo. 18s.

Mill.—*ANALYSIS OF THE PHENOMENA OF THE HUMAN MIND.* By JAMES MILL. With Notes, Illustrative and Critical. 2 vols. 8vo. 28s.

Mill.—*WORKS BY JOHN STUART MILL.*

PRINCIPLES OF POLITICAL ECONOMY. Library Edition, 2 vols. 8vo. 30s. People's Edition, 1 vol. crown 8vo. 5s.

A SYSTEM OF LOGIC, Ratiocinative and Inductive. Crown 8vo. 5s.

ON LIBERTY. Crown 8vo. 1s. 4d.

ON REPRESENTATIVE GOVERNMENT. Crown 8vo. 2s.

UTILITARIANISM. 8vo. 5s.

EXAMINATION OF SIR WILLIAM HAMILTON'S PHILOSOPHY. 8vo. 16s.

NATURE, THE UTILITY OF RELIGION, AND THEISM. Three Essays. 8vo. 5s.

Miller.—*WORKS BY W. ALLEN MILLER, M.D. LL.D.*

THE ELEMENTS OF CHEMISTRY, Theoretical and Practical. Re-edited, with Additions, by H. MACLEOD, F.C.S. 3 vols. 8vo.
Vol. I. CHEMICAL PHYSICS, 16s.
Vol. II. INORGANIC CHEMISTRY, 24s.
Vol. III. ORGANIC CHEMISTRY, 31s. 6d.

AN INTRODUCTION TO THE STUDY OF INORGANIC CHEMISTRY. With 71 Woodcuts. Fcp. 8vo. 3s. 6d.

Mitchell.—*A MANUAL OF PRACTICAL ASSAYING.* By JOHN MITCHELL, F.C.S. Revised, with the Recent Discoveries incorporated. By W. CROOKES, F.R.S. 8vo. Woodcuts, 31s. 6d.

Molesworth. — *MARRYING AND GIVING IN MARRIAGE:* a Novel. By Mrs. MOLESWORTH. Fcp. 8vo. 2s. 6d.

Monsell.—*WORKS BY THE REV. J. S. B. MONSELL, LL.D.*

SPIRITUAL SONGS FOR THE SUNDAYS AND HOLYDAYS THROUGHOUT THE YEAR. Fcp. 8vo. 5s. 18mo. 2s.

THE BEATITUDES. Eight Sermons. Crown 8vo. 3s. 6d.

HIS PRESENCE NOT HIS MEMORY. Verses. 16mo. 1s.

Mulhall.—*HISTORY OF PRICES SINCE THE YEAR 1850.* By MICHAEL G. MULHALL. Crown 8vo. 6s.

Munk.—*EUTHANASIA;* or, Medical Treatment in Aid of an Easy Death. By WILLIAM MUNK, M.D. F.S.A. Fellow and late Senior Censor of the Royal College of Physicians, &c. Crown 8vo. 4s. 6d.

Murchison.—*WORKS BY CHARLES MURCHISON, M.D. LL.D. &c.*

A TREATISE ON THE CONTINUED FEVERS OF GREAT BRITAIN. Revised by W. CAYLEY, M.D. Physician to the Middlesex Hospital. 8vo. with numerous Illustrations, 25s.

CLINICAL LECTURES ON DISEASES OF THE LIVER, JAUNDICE, AND ABDOMINAL DROPSY. Revised by T. LAUDER BRUNTON, M.D. and Sir JOSEPH FAYRER, M.D. 8vo. with 43 Illustrations, 24s.

Napier.—*THE LIFE OF SIR JOSEPH NAPIER, BART. EX-LORD CHANCELLOR OF IRELAND.* From his Private Correspondence. By ALEX. CHARLES EWALD, F.S.A. With Portrait on Steel, engraved by G. J. Stodart, from a Photograph. 8vo. 15s.

Nelson.—*LETTERS AND DESPATCHES OF HORATIO, VISCOUNT NELSON.* Selected and arranged by JOHN KNOX LAUGHTON, M.A. 8vo. 16s.

Nesbit.—*LAYS AND LEGENDS.* By E. NESBIT. Crown 8vo. 5s.

Newman.—*WORKS BY CARDINAL NEWMAN.*

APOLOGIA PRO VITÂ SÛA. Crown 8vo. 6s.

THE IDEA OF A UNIVERSITY DEFINED AND ILLUSTRATED. Crown 8vo. 7s.

HISTORICAL SKETCHES. 3 vols. crown 8vo. 6s. each.

THE ARIANS OF THE FOURTH CENTURY. Crown 8vo. 6s.

DISCUSSIONS AND ARGUMENTS ON VARIOUS SUBJECTS. Crown 8vo. 6s.

AN ESSAY ON THE DEVELOPMENT OF CHRISTIAN DOCTRINE. Crown 8vo. 6s.

CERTAIN DIFFICULTIES FELT BY ANGLICANS IN CATHOLIC TEACHING CONSIDERED. Vol. 1, crown 8vo. 7s. 6d.; Vol. 2, crown 8vo. 5s. 6d.

THE VIA MEDIA OF THE ANGLICAN CHURCH, ILLUSTRATED IN LECTURES &c. 2 vols. crown 8vo. 6s. each.

ESSAYS, CRITICAL AND HISTORICAL. 2 vols. crown 8vo. 12s.

ESSAYS ON BIBLICAL AND ON ECCLESIASTICAL MIRACLES. Crown 8vo. 6s.

AN ESSAY IN AID OF A GRAMMAR OF ASSENT. 7s. 6d.

THE DREAM OF GERONTIUS. 16mo. 6d. sewed.

Noble.—*HOURS WITH A THREE-INCH TELESCOPE.* By Captain W. NOBLE, F.R.A.S. &c. With a Map of the Moon. Crown 8vo. 4s. 6d.

Northcott.—*LATHES AND TURNING*, Simple, Mechanical, and Ornamental. By W. H. NORTHCOTT. With 338 Illustrations. 8vo. 18s.

O'Hagan.—*SELECTED SPEECHES AND ARGUMENTS OF THE RIGHT HONOURABLE THOMAS BARON O'HAGAN.* Edited by GEORGE TEELING. With a Portrait. 8vo. 16s.

Oliphant.—*NOVELS BY MRS. OLIPHANT.*

MADAM. Crown 8vo. 1s. boards; 1s. 6d. cloth.

IN TRUST.—Crown 8vo. 1s. boards; 1s. 6d. cloth.

Oliver. — *ASTRONOMY FOR TEURS:* a Practical Manual of Tel Research adapted to Moderate ments. Edited by J. A. WESTW OLIVER, with the assistance of E. MAUNDER, H. GRUBB, J. E. G(W. F. DENNING, W. S. FRANKS, T ELGER, S. W. BURNHAM, J. R. CAPI T. W. BACKHOUSE, and others. several Illustrations. Crown 8vo. 7s.

Overton.—*LIFE IN THE ENGLI CHURCH* (1660-1714). By J. H. OV TON, M.A. Rector of Epworth. 8vo. 1

Owen. — *THE COMPARATIVE A TOMY AND PHYSIOLOGY OF T VERTEBRATE ANIMALS.* By RICHARD OWEN, K.C.B. &c. With 1,(Woodcuts. 3 vols. 8vo. £3. 13s. 6d.

Paget. — *WORKS BY SIR JAM PAGET, BART. F.R.S. D.C.L.*

CLINICAL LECTURES AND ESS(Edited by F. HOWARD MARSH, Assis Surgeon to St. Bartholomew's Hosp: 8vo. 15s.

LECTURES ON SURGICAL PAT/ LOGY. Re-edited by the AUTHOR W. TURNER, M.B. 8vo. with 1 Woodcuts, 21s.

Pasteur.—*LOUIS PASTEUR, his L: and Labours.* By his SON-IN-LA Translated from the French by La CLAUD HAMILTON. Crown 8vo. 7s.

Payen.—*INDUSTRIAL CHEMISTR* a Manual for Manufacturers and for C(leges or Technical Schools; a Translati of PAYEN'S 'Précis de Chimie Indu trielle.' Edited by B. H. PAUL. WI 698 Woodcuts. Medium 8vo. 42s.

Payn.—*NOVELS BY JAMES PAYN.*

THE LUCK OF THE DARRELLS. Cro 8vo. 1s. boards; 1s. 6d. cloth.

THICKER THAN WATER. Crown 8v 1s. boards; 1s. 6d. cloth.

Pears.—*THE FALL OF CONSTANT NOPLE:* being the Story of the Fourt Crusade. By EDWIN PEARS, LL.[Barrister-at-Law, late President of th European Bar at Constantinople, an Knight of the Greek Order of th Saviour. 8vo. 16s.

Pennell.—*OUR SENTIMENTAL JOUR NEY THROUGH FRANCE AND ITAL* By JOSEPH and ELIZABETH ROBIN PENNELL. With a Map and 120 Illus trations by Joseph Pennell. Crown 8vo 6s. cloth or vegetable vellum.

Perring.—*HARD KNOTS IN SHAKE SPEARE.* By Sir PHILIP PERRING, Bart. 8vo. 7s. 6d.

Piesse.—*THE ART OF PERFUMERY,* and the Methods of Obtaining the Odours of Plants; with Instructions for the Manufacture of Perfumes, &c. By G. W. S. PIESSE, Ph.D. F.C.S. With 96 Woodcuts, square crown 8vo. 21*s.*

Pole.—*THE THEORY OF THE MODERN SCIENTIFIC GAME OF WHIST.* By W. POLE, F.R.S. Fcp. 8vo. 2*s.* 6*d.*

Prendergast.—*IRELAND,* from the Restoration to the Revolution, 1660–1690. By JOHN P. PRENDERGAST. 8vo. 5*s.*

Proctor.—*WORKS BY R. A. PROCTOR.*

THE ORBS AROUND US; a Series of Essays on the Moon and Planets, Meteors and Comets. With Chart and Diagrams, crown 8vo. 5*s.*

OTHER WORLDS THAN OURS; The Plurality of Worlds Studied under the Light of Recent Scientific Researches. With 14 Illustrations, crown 8vo. 5*s.*

THE MOON; her Motions, Aspects, Scenery, and Physical Condition. With Plates, Charts, Woodcuts, and Lunar Photographs, crown 8vo. 6*s.*

UNIVERSE OF STARS; Presenting Researches into and New Views respecting the Constitution of the Heavens. With 22 Charts and 22 Diagrams, 8vo. 10*s.* 6*d.*

LARGER STAR ATLAS for the Library, in 12 Circular Maps, with Introduction and 2 Index Pages. Folio, 15*s.* or Maps only, 12*s.* 6*d.*

NEW STAR ATLAS for the Library, the School, and the Observatory, in 12 Circular Maps (with 2 Index Plates). Crown 8vo. 5*s.*

LIGHT SCIENCE FOR LEISURE HOURS; Familiar Essays on Scientific Subjects, Natural Phenomena, &c. 3 vols. crown 8vo. 5*s.* each.

CHANCE AND LUCK; a Discussion of the Laws of Luck, Coincidences, Wagers, Lotteries, and the Fallacies of Gambling &c. Crown 8vo. 5*s.*

STUDIES OF VENUS-TRANSITS; an Investigation of the Circumstances of the Transits of Venus in 1874 and 1882. With 7 Diagrams and 10 Plates. 8vo. 5*s.*

OLD AND NEW ASTRONOMY.

⁎⁎* In course of publication, in twelve monthly parts and a supplementary section. In each there will be 64 pages, imp. 8vo. many cuts, and 2 plates, or one large folding plate. The price of each part will be 2*s.* 6*d.*; that of the supplementary section, containing tables, index, and preface, 1*s.* The price of the complete work, in cloth, 36*s.*

The 'KNOWLEDGE' LIBRARY. Edited by RICHARD A. PROCTOR.

HOW TO PLAY WHIST: WITH THE LAWS AND ETIQUETTE OF WHIST. By R. A. PROCTOR. Crown 8vo. 5*s.*

HOME WHIST: an Easy Guide to Correct Play. By R. A. PROCTOR. 16mo. 1*s.*

THE POETRY OF ASTRONOMY. A Series of Familiar Essays. By R. A. PROCTOR. Crown 8vo. 6*s.*

NATURE STUDIES. By GRANT ALLEN, A. WILSON, T. FOSTER, E. CLODD, and R. A. PROCTOR. Crown 8vo. 6*s.*

LEISURE READINGS. By E. CLODD, A. WILSON, T. FOSTER, A. C. RUNYARD, and R. A. PROCTOR. Crown 8vo. 6*s.*

THE STARS IN THEIR SEASONS. An Easy Guide to a Knowledge of the Star Groups, in 12 Large Maps. By R. A. PROCTOR. Imperial 8vo. 5*s.*

STAR PRIMER. Showing the Starry Sky Week by Week, in 24 Hourly Maps. By R. A. PROCTOR. Crown 4to. 2*s.* 6*d.*

THE SEASONS PICTURED IN 48 SUN-VIEWS OF THE EARTH, and 24 Zodiacal Maps, &c. By R. A. PROCTOR. Demy 4to. 5*s.*

STRENGTH AND HAPPINESS. By R. A. PROCTOR. Crown 8vo. 5*s.*

ROUGH WAYS MADE SMOOTH. Familiar Essays on Scientific Subjects. By R. A. PROCTOR. Crown 8vo. 5*s.*

OUR PLACE AMONG INFINITIES. A Series of Essays contrasting our Little Abode in Space and Time with the Infinities Around us. By R. A. PROCTOR. Crown 8vo. 5*s.*

THE EXPANSE OF HEAVEN. Essays on the Wonders of the Firmament. By R. A. PROCTOR. Crown 8vo. 5*s.*

THE GREAT PYRAMID, OBSERVATORY TOMB, AND TEMPLE. With Illustrations. Crown 8vo. 6*s.*

PLEASANT WAYS IN SCIENCE. By R. A. PROCTOR. Crown 8vo. 6*s.*

MYTHS AND MARVELS OF ASTRONOMY. By R. A. PROCTOR. Cr. 8vo. 6*s.*

Prothero.—*THE PIONEERS AND PROGRESS OF ENGLISH FARMING.* By ROWLAND E. PROTHERO. Crown 8vo. 5*s.*

Pryce.—*THE ANCIENT BRITISH CHURCH:* an Historical Essay. By JOHN PRYCE, M.A. Canon of Bangor. Crown 8vo. 6*s.*

Quain's Elements of Anatomy.
The Ninth Edition. Re-edited by ALLEN THOMSON, M.D. LL.D. F.R.S.S. L. & E. EDWARD ALBERT SCHÄFER, F.R.S. and GEORGE DANCER THANE. With upwards of 1,000 Illustrations engraved on Wood, of which many are Coloured. 2 vols. 8vo. 18s. each.

Quain.—*A DICTIONARY OF MEDICINE.* By Various Writers. Edited by R. QUAIN, M.D. F.R.S. &c. With 138 Woodcuts. Medium 8vo. 31s. 6d. cloth, or 40s. half-russia; to be had also in 2 vols. 34s. cloth.

Reader.—WORKS BY EMILY E. READER.

THE GHOST OF BRANKINSHAW and other Tales. With 9 Full-page Illustrations. Fcp. 8vo. 2s. 6d. cloth extra, gilt edges.

VOICES FROM FLOWER-LAND, in Original Couplets. A Birthday-Book and Language of Flowers. 16mo. 1s. 6d. limp cloth; 2s. 6d. roan, gilt edges, or in vegetable vellum, gilt top.

FAIRY PRINCE FOLLOW-MY-LEAD; or, the *MAGIC BRACELET.* Illustrated by WM. READER. Crown 8vo. 2s. 6d. gilt edges; or 3s. 6d. vegetable vellum, gilt edges.

THE THREE GIANTS &c. Royal 16mo. 1s. cloth.

THE MODEL BOY &c. Royal 16mo. 1s. cloth.

BE YT HYS WHO FYNDS YT. Royal 16mo. 1s. cloth.

Reeve. — *COOKERY AND HOUSEKEEPING.* By Mrs. HENRY REEVE. With 8 Coloured Plates and 37 Woodcuts. Crown 8vo. 5s.

Rich.—*A DICTIONARY OF ROMAN AND GREEK ANTIQUITIES.* With 2,000 Woodcuts. By A. RICH, B.A. Cr. 8vo. 7s. 6d.

Richardson.—WORKS BY BENJAMIN WARD RICHARDSON, M.D.

THE HEALTH OF NATIONS: a Review of the Works—Economical, Educational, Sanitary, and Administrative—of EDWIN CHADWICK, C.B. With a Biographical Dissertation by BENJAMIN WARD RICHARDSON, M.D. F.R.S. 2 vols. 8vo. 28s.

THE COMMONHEALTH: a Series of Essays on Health and Felicity for EveryDay Readers. Crown 8vo. 6s.

Richey.—*A SHORT HISTORY OF T. IRISH PEOPLE*, down to the Date of Plantation of Ulster. By the late A. RICHEY, Q.C. LL.D. M.R.I.A. Edit with Notes, by ROBERT ROMNEY KAN LL.D. M.R.I.A. 8vo. 14s.

Riley.—*ATHOS;* or, the Mountain the Monks. By ATHELSTAN RILE M.A. F.R.G.S. With Map and 2 Illustrations. 8vo. 21s.

Rivers. — WORKS BY THOMA RIVERS.

THE ORCHARD-HOUSE. With 2 Woodcuts. Crown 8vo. 5s.

THE MINIATURE FRUIT GARDEN or, the Culture of Pyramidal and Fruit Trees, with Instructions for R Pruning. With 32 Illustrations. F 8vo. 4s.

Roberts.— GREEK THE LANGUA(OF CHRIST AND HIS APOSTLES. ALEXANDER ROBERTS, D.D. 8vo. 1

Robinson. — *THE NEW ARCADI* and other Poems. By A. MARY ROBINSON. Crown 8vo. 6s.

Roget.— *THESAURUS OF ENGLI? WORDS AND PHRASES*, Classified a Arranged so as to facilitate the Expressi of Ideas and assist in Literary position. By PETER M. ROGET. (8vo. 10s. 6d.

Ronalds. — *THE FLY-FISH ENTOMOLOGY.* By ALFRED RON With 20 Coloured Plates. 8vo. 14s.

Saintsbury.—*MANCHESTER:* a Sho History. By GEORGE SAINTSBURY. Wi 2 Maps. Crown 8vo. 3s. 6d.

Schäfer. — *THE ESSENTIALS (HISTOLOGY, DESCRIPTIVE AND PRAC1 CAL.* For the use of Students. By A. SCHÄFER, F.R.S. With 281 Ill trations. 8vo. 6s. or Interleaved wi Drawing Paper, 8s. 6d.

Schellen. — *SPECTRUM ANALYS: IN ITS APPLICATION TO TERRESTRI. SUBSTANCES*, and the Physical Consti tion of the Heavenly Bodies. By I H. SCHELLEN. Translated by JANE a CAROLINE LASSELL. Edited by Ca W. DE W. ABNEY. With 14 Pla (including Angström's and Cornu's Ma and 291 Woodcuts. 8vo. 31s. 6d.

Scott.—*WEATHER CHARTS A STORM WARNINGS.* By ROBERT SCOTT, M.A. F.R.S. With numero Illustrations. Crown 8vo. 6s.

Seebohm.—*WORKS BY FREDERIC SEEBOHM.*

THE OXFORD REFORMERS — JOHN COLET, ERASMUS, AND THOMAS MORE; a History of their Fellow-Work. 8vo. 14s.

THE ENGLISH VILLAGE COMMUNITY Examined in its Relations to the Manorial and Tribal Systems, &c. 13 Maps and Plates. 8vo. 16s.

THE ERA OF THE PROTESTANT REVOLUTION. With Map. Fcp. 8vo. 2s. 6d.

Sennett.—*THE MARINE STEAM ENGINE;* a Treatise for the use of Engineering Students and Officers of the Royal Navy. By RICHARD SENNETT, Engineer-in-Chief of the Royal Navy. With 244 Illustrations. 8vo. 21s.

Sewell. — *STORIES AND TALES.* By ELIZABETH M. SEWELL. Crown 8vo. 1s. each, boards; 1s. 6d. each, cloth plain; 2s. 6d. each, cloth extra, gilt edges:—

Amy Herbert.	Margaret Percival.
The Earl's Daughter.	Laneton Parsonage.
The Experience of Life.	Ursula.
A Glimpse of the World.	Gertrude.
Cleve Hall.	Ivors.
Katharine Ashton.	

Shakespeare. — *BOWDLER'S FAMILY SHAKESPEARE.* Genuine Edition, in 1 vol. medium 8vo. large type, with 36 Woodcuts, 14s. or in 6 vols. fcp. 8vo. 21s.

OUTLINES OF THE LIFE OF SHAKESPEARE. By J. O. HALLIWELL-PHILLIPPS, F.R.S. 2 vols. Royal 8vo. 10s. 6d.

Shilling Standard Novels.

BY THE EARL OF BEACONSFIELD.

Vivian Grey.	The Young Duke, &c.
Venetia.	Contarini Fleming,&c.
Tancred.	Henrietta Temple.
Sybil.	Lothair.
Coningsby.	Endymion.
Alroy, Ixion, &c.	

Price 1s. each, boards; 1s. 6d. each, cloth.

BY G. J. WHYTE-MELVILLE.

The Gladiators.	Holmby House.
The Interpreter.	Kate Coventry.
Good for Nothing.	Digby Grand.
Queen's Maries.	General Bounce.

Price 1s. each, boards; 1s. 6d. each, cloth.

BY ROBERT LOUIS STEVENSON.

The Dynamiter.
Strange Case of Dr. Jekyll and Mr. Hyde.
Price 1s. each, sewed; 1s. 6d. each, cloth.

[Continued above.

Shilling Standard Novels—*contd.*

BY ELIZABETH M. SEWELL.

Amy Herbert.	A Glimpse of the World.
Gertrude.	Ivors.
Earl's Daughter.	Katharine Ashton.
The Experience of Life.	Margaret Percival.
	Laneton Parsonage.
Cleve Hall.	Ursula.

Price 1s. each, boards; 1s. 6d. each, cloth, plain; 2s. 6d. each, cloth extra, gilt edges.

BY ANTHONY TROLLOPE.

The Warden. | Barchester Towers.
Price 1s. each, boards; 1s. 6d. each, cloth.

BY BRET HARTE.

In the Carquinez Woods. 1s. boards; 1s. 6d. cloth.
On the Frontier (Three Stories). 1s. sewed.
By Shore and Sedge (Three Stories). 1s. sewed.

BY MRS. OLIPHANT.

In Trust. | Madam.

BY JAMES PAYN.

Thicker than Water.
The Luck of the Darrells.
Price 1s. each, boards; 1s. 6d. each, cloth.

Short.—*SKETCH OF THE HISTORY OF THE CHURCH OF ENGLAND TO THE REVOLUTION OF* 1688. By T. V. SHORT, D.D. Crown 8vo. 7s. 6d.

Smith.—*LIBERTY AND LIBERALISM;* a Protest against the Growing Tendency toward Undue Interference by the State with Individual Liberty, Private Enterprise, and the Rights of Property. By BRUCE SMITH, of the Inner Temple, Barrister-at-Law. Crown 8vo. 6s.

Smith, H. F.—*THE HANDBOOK FOR MIDWIVES.* By HENRY FLY SMITH, M.B. Oxon. M.R.C.S. late Assistant-Surgeon at the Hospital for Sick Women, Soho Square. With 41 Woodcuts. Crown 8vo. 5s.

Smith, R. Bosworth. — *CARTHAGE AND THE CARTHAGINIANS.* By R. BOSWORTH SMITH, M.A. Maps, Plans, &c. Crown 8vo. 10s. 6d.

Smith, Rev. Sydney.—*THE WIT AND WISDOM OF THE REV. SYDNEY SMITH.* Crown 8vo. 1s. boards; 1s. 6d. cloth.

Smith, T.—*A MANUAL OF OPERATIVE SURGERY ON THE DEAD BODY.* By THOMAS SMITH, Surgeon to St. Bartholomew's Hospital. A New Edition, re-edited by W. J. WALSHAM. With 46 Illustrations. 8vo. 12s.

Southey.—THE POETICAL WORKS OF ROBERT SOUTHEY, with the Author's last Corrections and Additions. Medium 8vo. with Portrait, 14s.

Stanley. — A FAMILIAR HISTORY OF BIRDS. By E. STANLEY, D.D. Revised and enlarged, with 160 Woodcuts. Crown 8vo. 6s.

Steel.—WORKS BY J. H. STEEL, M.R.C.V.S.

A TREATISE ON THE DISEASES OF THE DOG; being a Manual of Canine Pathology. Especially adapted for the Use of Veterinary Practitioners and Students. With 88 Illustrations. 8vo. 10s. 6d.

A TREATISE ON THE DISEASES OF THE OX; being a Manual of Bovine Pathology specially adapted for the use of Veterinary Practitioners and Students. With 2 Plates and 117 Woodcuts. 8vo. 15s.

Stephen. — ESSAYS IN ECCLESIASTICAL BIOGRAPHY. By the Right Hon. Sir J. STEPHEN, LL.D. Crown 8vo. 7s. 6d.

Stevenson.—WORKS BY ROBERT LOUIS STEVENSON.

A CHILD'S GARDEN OF VERSES. Small fcp. 8vo. 5s.

THE DYNAMITER. Fcp. 8vo. 1s. swd. 1s. 6d. cloth.

STRANGE CASE OF DR. JEKYLL AND MR. HYDE. Fcp. 8vo. 1s. sewed; 1s. 6d. cloth.

'Stonehenge.' — THE DOG IN HEALTH AND DISEASE. By 'STONEHENGE.' With 84 Wood Engravings. Square crown 8vo. 7s. 6d.

THE GREYHOUND. By 'STONEHENGE.' With 25 Portraits of Greyhounds, &c. Square crown 8vo. 15s.

Stoney. — THE THEORY OF THE STRESSES ON GIRDERS AND SIMILAR STRUCTURES. With Practical Observations on the Strength and other Properties of Materials. By BINDON B. STONEY, LL.D. F.R.S. M.I.C.E. With 5 Plates, and 143 Illustrations in the Text. Royal 8vo. 36s.

Sully.—WORKS BY JAMES SULLY.

OUTLINES OF PSYCHOLOGY, wi Special Reference to the Theory of Ed cation. 8vo. 12s. 6d.

THE TEACHER'S HANDBOOK (PSYCHOLOGY, on the Basis of 'Ou of Psychology.' Crown 8vo. 6s. 6d.

Supernatural Religion; an quiry into the Reality of Divine R lation. Complete Edition, thoroug revised. 3 vols. 8vo. 36s.

Swinburne. — PICTURE LOGIC; Attempt to Popularise the Science Reasoning. By A. J. SWINBURNE, B. Post 8vo. 5s.

Taylor. — STUDENT'S MANUAL (THE HISTORY OF INDIA, from the Earli Period to the Present Time. By Col MEADOWS TAYLOR, C.S.I. Crown 8 7s. 6d.

Taylor.—AN AGRICULTURAL NOT BOOK: to Assist Candidates in P paring for the Science and Art and oth Examinations in Agriculture. By W. TAYLOR. Crown 8vo. 2s. 6d.

Thompson.—WORKS BY D. GREE LEAF THOMPSON.

THE PROBLEM OF EVIL: an Intr duction to the Practical Sciences. 8 10s. 6d.

A SYSTEM OF PSYCHOLOGY. 2 VO 8vo. 36s.

Thomson's Conspectus.—Adapt to the British Pharmacopœia of 18 Edited by NESTOR TIRARD, M.D. Lon F.R.C.P. New Edition, with an A pendix containing notices of some of t more important non-official medicin and preparations. 18mo. 6s.

Thomson.—AN OUTLINE OF T NECESSARY LAWS OF THOUGHT; Treatise on Pure and Applied Logic. W. THOMSON, D.D. Archbishop York. Crown 8vo. 6s.

Three in Norway. By Two THEM. With a Map and 59 Illus tions on Wood from Sketches by Authors. Crown 8vo. 2s. boards; 2s. cloth.

Todd. — *ON PARLIAMENTARY GO-VERNMENT IN ENGLAND :* its Origin, Development, and Practical Operation. By ALPHEUS TODD, LL.D. C.M.G. Librarian of Parliament for the Dominion of Canada. Second Edition, by his SON. In Two Volumes—VOL. I. 8vo. 24*s.*

Trevelyan. — *WORKS BY THE RIGHT HON. SIR G. O. TREVELYAN, BART.*

THE LIFE AND LETTERS OF LORD MACAULAY.

LIBRARY EDITION, 2 vols. 8vo. 36*s.*
CABINET EDITION, 2 vols. crown 8vo. 12*s.*
POPULAR EDITION, 1 vol. crown 8vo. 6*s.*

THE EARLY HISTORY OF CHARLES JAMES FOX. Library Edition, 8vo. 18*s.* Cabinet Edition, crown 8vo. 6*s.*

Trollope. — *NOVELS BY ANTHONY TROLLOPE.*

THE WARDEN. Crown 8vo. 1*s.* boards ; 1*s. 6d.* cloth.
BARCHESTER TOWERS. Crown 8vo. 1*s.* boards ; 1*s. 6d.* cloth.

Twiss. — *WORKS BY SIR TRAVERS TWISS.*

THE RIGHTS AND DUTIES OF NA-TIONS, considered as Independent Com-munities in Time of War. 8vo. 21*s.*
THE RIGHTS AND DUTIES OF NATIONS IN TIME OF PEACE. 8vo. 15*s.*

Tyndall. — *WORKS BY JOHN TYN-DALL, F.R.S. &c.*

FRAGMENTS OF SCIENCE. 2 vols. crown 8vo. 16*s.*
HEAT A MODE OF MOTION. Crown 8vo. 12*s.*
SOUND. With 204 Woodcuts. Crown 8vo. 10*s. 6d.*
ESSAYS ON THE FLOATING-MATTER OF THE AIR in relation to Putrefaction and Infection. With 24 Woodcuts. Crown 8vo. 7*s. 6d.*
LECTURES ON LIGHT, delivered in America in 1872 and 1873. With 57 Diagrams. Crown 8vo. 5*s.*
LESSONS IN ELECTRICITY AT THE ROYAL INSTITUTION, 1875-76. With 58 Woodcuts. Crown 8vo. 2*s. 6d.*
NOTES OF A COURSE OF SEVEN LECTURES ON ELECTRICAL PHENO-MENA AND THEORIES, delivered at the Royal Institution. Crown 8vo. 1*s.* sewed, 1*s. 6d.* cloth.

[*Continued above.*

Tyndall. — *WORKS BY JOHN TYN-DALL, F.R.S. &c.—continued.*

NOTES OF A COURSE OF NINE LEC-TURES ON LIGHT, delivered at the Royal Institution. Crown 8vo. 1*s.* sewed, 1*s. 6d.* cloth.
FARADAY AS A DISCOVERER. Fcp. 8vo. 3*s. 6d.*

Ville. — *ON ARTIFICIAL MANURES,* their Chemical Selection and Scientific Application to Agriculture. By GEORGES VILLE. Translated and edited by W. CROOKES, F.R.S. With 31 Plates. 8vo. 21*s.*

Virgil. — *PUBLI VERGILI MARONIS BUCOLICA, GEORGICA, ÆNEIS ;* the Works of VIRGIL, Latin Text, with English Commentary and Index. By B. H. KENNEDY, D.D. Crown 8vo. 10*s. 6d.*

THE ÆNEID OF VIRGIL. Translated into English Verse. By JOHN CONING-TON, M.A. Crown 8vo. 9*s.*
THE POEMS OF VIRGIL. Translated into English Prose. By JOHN CONING-TON, M.A. Crown 8vo. 9*s.*

Vitzthum. — *ST. PETERSBURG AND LONDON IN THE YEARS 1852-1864 :* Reminiscences of Count CHARLES FRED-ERICK VITZTHUM VON ECKSTOEDT, late Saxon Minister at the Court of St. James'. Edited, with a Preface, by HENRY REEVE, C.B. D.C.L. 2 vols. 8vo. 30*s.*

Walker. — *THE CORRECT CARD ;* or, How to Play at Whist ; a Whist Catechism. By Major A. CAMPBELL-WALKER, F.R.G.S. Fcp. 8vo. 2*s. 6d.*

Walpole. — *HISTORY OF ENGLAND FROM THE CONCLUSION OF THE GREAT WAR IN 1815.* By SPENCER WALPOLE. 5 vols. 8vo. Vols. I. and II. 1815-1832, 36*s.* ; Vol. III. 1832-1841, 18*s.* ; Vols. IV. and V. 1841-1858, 36*s.*

Waters. — *PARISH REGISTERS IN ENGLAND :* their History and Contents. With Suggestions for Securing their better Custody and Preservation. By ROBERT E. CHESTER WATERS, B.A. 8vo. 5*s.*

Watson. — *MARAHUNA :* a Romance. By H. B. MARRIOTT WATSON. Crown 8vo. 6*s.*

Watts. — *A DICTIONARY OF CHEMIS-TRY AND THE ALLIED BRANCHES OF OTHER SCIENCES.* Edited by HENRY WATTS, F.R.S. 9 vols. medium 8vo. £15. 2*s. 6d.*

Webb.—*CELESTIAL OBJECTS FOR COMMON TELESCOPES.* By the Rev. T. W. WEBB. Map, Plate, Woodcuts. Crown 8vo. 9s.

Wellington.—*LIFE OF THE DUKE OF WELLINGTON.* By the Rev. G. R. GLEIG, M.A. Crown 8vo. Portrait, 6s.

West.—*WORKS BY CHARLES WEST, M.D. &c.* Founder of, and formerly Physician to, the Hospital for Sick Children.

LECTURES ON THE DISEASES OF INFANCY AND CHILDHOOD. 8vo. 18s.

THE MOTHER'S MANUAL OF CHILDREN'S DISEASES. Crown 8vo. 2s. 6d.

Whately. — *ENGLISH SYNONYMS.* By E. JANE WHATELY. Edited by her Father, R. WHATELY, D.D. Fcp. 8vo. 3s.

Whately.—*WORKS BY R. WHATELY, D.D.*

ELEMENTS OF LOGIC. Crown 8vo. 4s. 6d.

ELEMENTS OF RHETORIC. Crown 8vo. 4s. 6d.

LESSONS ON REASONING. Fcp. 8vo. 1s. 6d.

BACON'S ESSAYS, with Annotations. 8vo. 10s. 6d.

White and Riddle.—*A LATIN-ENGLISH DICTIONARY.* By J. T. WHITE, D.D. Oxon. and J. J. E. RIDDLE, M.A. Oxon. Founded on the larger Dictionary of Freund. Royal 8vo. 21s.

White.—*A CONCISE LATIN-ENGLISH DICTIONARY*, for the Use of Advanced Scholars and University Students By the Rev. J. T. WHITE, D.D. Royal 8vo. 12s.

Whiteing.—*THE ISLAND:* an Adventure of a Person of Quality; a Novel. By RICHARD WHITEING. Crown 8vo. 6s.

Wilcocks.—*THE SEA FISHERMAN.* Comprising the Chief Methods of Hook and Line Fishing in the British and other Seas, and Remarks on Nets, Boats, and Boating. By J. C. WILCOCKS. Profusely Illustrated. Crown 8vo. 6s.

Wilkinson.—*THE FRIENDLY SOCIETY MOVEMENT:* Its Origin, Rise, and Growth; its Social, Moral, and Educational Influences.—*THE AFFILIATED ORDERS.* —By the Rev. JOHN FROME WILKINSON, M.A. Crown 8vo. 2s. 6d.

Williams.—*PULMONARY CONSUTION;* its Etiology, Pathology, Treatment. With an Analysis of 1, Cases to Exemplify its Duration Modes of Arrest. By C. J. B. WILLIA M.D. LL.D. F.R.S. F.R.C.P. CHARLES THEODORE WILLIAMS, M. M.D.Oxon. F.R.C.P. With 4 Colou Plates and 10 Woodcuts. 8vo. 16s.

Williams. — *MANUAL OF TEl GRAPHY.* By W. WILLIAMS, Super tendent of Indian Government Telegrap Illustrated by 93 Wood Engravings. 8 10s. 6d.

Willich. — *POPULAR TABLES* giving Information for ascertaining value of Lifehold, Leasehold, and Chu Property, the Public Funds, &c. CHARLES M. WILLICH. Edited H. BENCE JONES. Crown 8vo. 10s.

Wilson.—*A MANUAL OF HEAL: SCIENCE.* Adapted for Use in Sch and Colleges, and suited to the Req ments of Students preparing for the aminations in Hygiene of the Scie and Art Department, &c. By ANDR WILSON, F.R.S.E. F.L.S. &c. Wi 74 Illustrations. Crown 8vo. 2s. 6d.

Witt.—*WORKS BY PROF. WIT* Translated from the German by FRANC YOUNGHUSBAND.

THE TROJAN WAR. With a Prefa by the Rev. W. G. RUTHERFORD, M. Head-Master of Westminster Sch Crown 8vo. 2s.

MYTHS OF HELLAS; or, Greek Tal Crown 8vo. 3s. 6d.

THE WANDERINGS OF ULYSS Crown 8vo. 3s. 6d.

Wood.—*WORKS BY REV. J. WOOD.*

HOMES WITHOUT HANDS; a scription of the Habitations of Ani classed according to the Principle of (struction. With 140 Illustrations. 10s. 6d.

INSECTS AT HOME; a Pop Account of British Insects, their S ture, Habits, and Transformations. 700 Illustrations. 8vo. 10s. 6d.

INSECTS ABROAD; a Popular Acc of Foreign Insects, their Struc Habits, and Transformations. 600 Illustrations. 8vo. 10s. 6d.

[Continued on next page

Wood. — *WORKS BY REV. J. G. WOOD—continued.*

BIBLE ANIMALS; a Description of every Living Creature mentioned in the Scriptures. With 112 Illustrations. 8vo. 10s. 6d.

STRANGE DWELLINGS; a Description of the Habitations of Animals, abridged from 'Homes without Hands.' With 60 Illustrations. Crown 8vo. 5s. Popular Edition, 4to. 6d.

HORSE AND MAN: their Mutual Dependence and Duties. With 49 Illustrations. 8vo. 14s.

ILLUSTRATED STABLE MAXIMS. To be hung in Stables for the use of Grooms, Stablemen, and others who are in charge of Horses. On Sheet, 4s.

OUT OF DOORS; a Selection of Original Articles on Practical Natural History. With 11 Illustrations. Crown 8vo. 5s.

PETLAND REVISITED. With 33 Illustrations. Crown 8vo. 7s. 6d.

The following books are extracted from the foregoing works by the Rev. J. G. WOOD :

SOCIAL HABITATIONS AND PARASITIC NESTS. With 18 Illustrations. Crown 8vo. 2s. cloth extra, gilt edges.

THE BRANCH BUILDERS. With 28 Illustrations. Crown 8vo. 2s. 6d. cloth extra, gilt edges.

WILD ANIMALS OF THE BIBLE. With 29 Illustrations. Crown 8vo. 3s. 6d. cloth extra, gilt edges.

DOMESTIC ANIMALS OF THE BIBLE. With 23 Illustrations. Crown 8vo. 3s. 6d. cloth extra, gilt edges.

BIRD-LIFE OF THE BIBLE. With 32 Illustrations. Crown 8vo. 3s. 6d. cloth extra, gilt edges.

WONDERFUL NESTS. With 30 Illustrations. Crown 8vo. 3s. 6d. cloth extra, gilt edges.

HOMES UNDER THE GROUND. With 28 Illustrations. Crown 8vo. 3s. 6d. cloth extra, gilt edges.

Wood-Martin. — *THE LAKE DWELLINGS OF IRELAND:* or Ancient Lacustrine Habitations of Erin, commonly called Crannogs. By W. G. WOOD-MARTIN, M.R.I.A. Lieut.-Colonel 8th Brigade North Irish Division, R.A. With 50 Plates. Royal 8vo. 25s.

Wright. — *HIP DISEASE IN CHILD-HOOD,* with Special Reference to its Treatment by Excision. By G. A. WRIGHT, B.A. M.B.Oxon. F.R.C.S.Eng. With 48 Original Woodcuts. 8vo. 10s. 6d.

Wylie. — *HISTORY OF ENGLAND UNDER HENRY THE FOURTH.* By JAMES HAMILTON WYLIE, M.A. one of Her Majesty's Inspectors of Schools. (2 vols.) Vol. 1, crown 8vo. 10s. 6d.

Wylie. — *LABOUR, LEISURE, AND LUXURY;* a Contribution to Present Practical Political Economy. By ALEXANDER WYLIE, of Glasgow. Crown 8vo. 1s.

Youatt. — *WORKS BY WILLIAM YOUATT.*

THE HORSE. Revised and enlarged by W. WATSON, M.R.C.V.S. 8vo. Woodcuts, 7s. 6d.

THE DOG. Revised and enlarged. 8vo. Woodcuts. 6s.

Younghusband. — *THE STORY OF OUR LORD, TOLD IN SIMPLE LANGUAGE FOR CHILDREN.* By FRANCES YOUNG-HUSBAND. With 25 Illustrations on Wood from Pictures by the Old Masters, and numerous Ornamental Borders, Initial Letters, &c. from Longmans' Illustrated New Testament. Crown 8vo. 2s. 6d. cloth plain ; 3s. 6d. cloth extra, gilt edges.

Zeller. — *WORKS BY DR. E. ZELLER.*

HISTORY OF ECLECTICISM IN GREEK PHILOSOPHY. Translated by SARAH F. ALLEYNE. Crown 8vo. 10s. 6d.

THE STOICS, EPICUREANS, AND SCEPTICS. Translated by the Rev. O. J. REICHEL, M.A. Crown 8vo. 15s.

SOCRATES AND THE SOCRATIC SCHOOLS. Translated by the Rev. O. J. REICHEL, M.A. Crown 8vo. 10s. 6d.

PLATO AND THE OLDER ACADEMY. Translated by SARAH F. ALLEYNE and ALFRED GOODWIN, B.A. Crown 8vo. 18s.

THE PRE-SOCRATIC SCHOOLS : a History of Greek Philosophy from the Earliest Period to the time of Socrates. Translated by SARAH F. ALLEYNE. 2 vols. crown 8vo. 30s.

OUTLINES OF THE HISTORY OF GREEK PHILOSOPHY. Translated by SARAH F. ALLEYNE and EVELYN ABBOTT. Crown 8vo. 10s. 6d.

EPOCHS OF ANCIENT HISTORY.

Edited by the Rev. Sir G. W. Cox, Bart. M.A. and by C. SANKEY, M.A.   10 volumes,
fcp. 8vo. with Maps, price 2s. 6d. each.

THE GRACCHI, MARIUS, AND SULLA. By
A. H. BEESLY, M.A. With 2 Maps.

THE EARLY ROMAN EMPIRE. From the
Assassination of Julius Cæsar to the Assassination
of Domitian. By the Rev. W. WOLFE CAPES, M.A.
With 2 Maps.

THE ROMAN EMPIRE OF THE SECOND CEN-
tury, or the Age of the Antonines. By the Rev.
W. WOLFE CAPES, M.A. With 2 Maps.

THE ATHENIAN EMPIRE FROM THE FLIGHT
of Xerxes to the Fall of Athens. By the Rev.
Sir G. W. Cox, Bart. M.A. With 5 Maps.

THE RISE OF THE MACEDONIAN EMPIRE.
By ARTHUR M. CURTEIS, M.A. With 8 Maps.

THE GREEKS AND THE PERSIANS. By
Rev. Sir G. W. Cox, Bart. M.A. With 4 Ma

ROME TO ITS CAPTURE BY THE GAU
By WILHELM IHNE. With a Map.

THE ROMAN TRIUMVIRATES. By the V
Rev. CHARLES MERIVALE, D.D. Dean of l
With a Map.

THE SPARTAN AND THEBAN SUPREMACI
By CHARLES SANKEY, M.A. With 5 Maps.

ROME AND CARTHAGE, THE PUNIC WA
By R. BOSWORTH SMITH, M.A. With 9 M
and Plans.

EPOCHS OF MODERN HISTORY.

Edited by C. COLBECK, M.A.   18 volumes, fcp. 8vo. with Maps, price 2s. 6d. each.

THE BEGINNING OF THE MIDDLE AGES.
By the Very Rev. RICHARD WILLIAM CHURCH,
M.A. &c. Dean of St. Paul's. With 3 Maps.

THE NORMANS IN EUROPE. By Rev. A.
H. JOHNSON, M.A. With 3 Maps.

THE CRUSADES. By the Rev. Sir G. W.
Cox, Bart. M.A. With a Map.

THE EARLY PLANTAGENETS. By the
Right Rev. W. STUBBS, D.D. Bishop of Chester.
With 2 Maps.

EDWARD THE THIRD. By the Rev. W.
WARBURTON, M.A. With 3 Maps and 3 Genealogical Tables.

THE HOUSES OF LANCASTER AND YORK;
with the Conquest and Loss of France. By
JAMES GAIRDNER. With 5 Maps.

THE EARLY TUDORS. By the Rev. C. E.
MOBERLY, M.A.

THE ERA OF THE PROTESTANT REVOLU-
tion. By F. SEEBOHM. With 4 Maps and 12
Diagrams.

THE AGE OF ELIZABETH. By the Rev.
CREIGHTON, M.A. LL.D.   With 5 Maps
4 Genealogical Tables.

THE FIRST TWO STUARTS AND THE PU
tan Revolution, 1603-1660. By SAMUEL RAW
GARDINER. With 4 Maps.

THE FALL OF THE STUARTS; AND WEST
Europe from 1678 to 1697. By the Rev. EDW
HALE, M.A. With 11 Maps and Plans.

THE AGE OF ANNE. By E. E. MOR
M.A. With 7 Maps and Plans.

THE THIRTY YEARS' WAR, 1618-1648.
SAMUEL RAWSON GARDINER. With a Map.

THE EARLY HANOVERIANS. By E.
MORRIS, M.A. With 9 Maps and Plans.

FREDERICK THE GREAT AND THE SE
Years' War. By F. W. LONGMAN. With 2 M

THE WAR OF AMERICAN INDEPEND
1775-1783. By J. M. LUDLOW. With 4 Ma

THE FRENCH REVOLUTION, 1789-1795.
Mrs. S. R. GARDINER. With 7 Maps.

THE EPOCH OF REFORM, 1830-1850.
JUSTIN MCCARTHY, M.P.

EPOCHS OF CHURCH HISTORY.

Edited by the Rev. MANDELL CREIGHTON. Fcp. 8vo. price 2s. 6d. each.

THE ENGLISH CHURCH IN OTHER LANDS.
By the Rev. H. W. TUCKER, M.A.

THE HISTORY OF THE REFORMATION IN
England. By the Rev. GEORGE G. PERRY, M.A.

THE CHURCH OF THE EARLY FATHERS.
By ALFRED PLUMMER, D.D.

THE EVANGELICAL REVIVAL IN THE
Eighteenth Century. By the Rev. J. H. OVER-
TON, M.A.

THE HISTORY OF THE UNIVERSITY
Oxford. By the Hon. G. C. BRODRICK, D.

THE CHURCH AND THE ROMAN EM
By the Rev. A. CARR.

THE CHURCH AND THE PURITANS, 1
1660. By HENRY OFFLEY WAKEMAN, M.

THE CHURCH AND THE EASTERN EMP
By the Rev. H. F. TOZER, M.A.

HILDEBRAND AND HIS TIMES. By
Rev. W. R. W. STEPHENS, M.A.

•*• Other Volumes are in preparation.

Spottiswoode & Co. Printers, New-street Square, London.

Lightning Source UK Ltd.
Milton Keynes UK
UKHW010613120219
337137UK00007B/1376/P